# GELS HANDBOOK

## Volume 4

# GELS HANDBOOK

## Volume 4

### Environment:
### Earth Environment & Gels

**Editors-in-Chief**

Yoshihito Osada and Kanji Kajiwara

**Associate Editors**

Takao Fushimi, Okihiko Hirasa,
Yoshitsugu Hirokawa, Tsutomu Matsunaga,
Tadao Shimomura, and Lin Wang

**Translated by**

Hatsuo Ishida

## ACADEMIC PRESS

A Harcourt Science and Technology Company

San Diego   San Francisco   New York   Boston
London   Sydney   Tokyo

ACADEMIC PRESS
*A Harcourt Science and Technology Company*
525 B Street, Suite 1900, San Diego, CA 92101-4495, USA
http://www.academicpress.com

Academic Press
Harcourt Place, 32 Jamestown Road, London, NW1 7BY, UK

**Library of Congress Catalog Number**: 00-107106
**International Standard Book Number**: 0-12-394690-5 (Set)

International Standard Book Number, Volume 4: 0-12-394964-5

Printed in the United States of America
00  01  02  03  04  IP  9  8  7  6  5  4  3  2  1

# Contents

## Chapter 2　Forestation Technology　21

## Chapter 3　Sanitary Products and Environmental Problems　21

## Data Summary   Gel compound data index   75

# *Preface*

The development, production, and application of superabsorbent gels is increasing at a remarkable pace. Research involving functional materials in such areas as medical care, medicine, foods, civil engineering, bioengineering, and sports is already widely documented. In the twenty-first century innovative research and development is growing ever more active. Gels are widely expected to be one of the essential solutions to various problems such as limited food resources, environmental preservation, and safeguarding human welfare.

In spite of the clear need for continued gel research and development, there have been no comprehensive references involving gels until now. In 1996, an editorial board led by the main members of the Association of Polymer Gel Research was organized with the primary goal of collecting a broad range of available information and organizing this information in such a way that would be helpful for not only gels scientists, but also for researchers and engineers in other fields. The content covers all topics ranging from preparation methods, structure, and characteristics to applications, functions, and evaluation methods of gels. It consists of Volume 1, The Fundamentals; Volume 2, Functions; Volume 3, Applications; and Volume 4, Environment: Earth Environment and Gels, which consists of several appendices and an index on gel compounds.

Because we were fortunate enough to receive contributions from the leading researchers on gels in Japan and abroad, we offer this book with great confidence. We would like to thank the editors as well as the authors who willingly contributed despite their very busy schedules.

This handbook was initially proposed by Mr. Shi Matsunaga. It is, of course, due to the neverending effort by him and the editorial staff that this handbook was successfully completed. We would also like to express great appreciation to the enthusiasm and help of Mr. Takashi Yoshida and Ms. Masami Matsukaze of NTS Inc.

Yoshihito Osada
Kanji Kajiwara
*November, 1997*

# *Contributors*

***Editors-in-Chief***

Yoshihito Osada, *Professor, Department of Scientific Research, Division of Biology at Hokkaido University Graduate School*

Kanji Kajiwara, *Professor, Department of Technical Art in Material Engineering at Kyoto University of Industrial Art and Textile*

***Principal Editorial Members***

Tadao Shimomura, *President, Japan Catalytic Polymer Molecule Research Center*

Okihiko Hirasa, *Professor, Department of Education and Domestic Science at Iwate University*

Yoshitsugu Hirokawa, *Technical Councilor, Science and Technology Promotional Office, Hashimoto Phase Separation Structure Project*

Takao Fushimi, *Examiner, Patent Office Third Examination Office at Ministry of International Trade and Industry*

Tsutomu Matsunaga, *Director, Chemistry Bio-Tsukuba*

Lin Wang, *Senior Scientist, P&G Product Development Headquarters*

Ito Takeshi, *Assistant Manager, Tokyo Office Sales and Development Division of Mitsubishi Chemical Co.*

Seigo Ouchi, *Head Researcher, Kanishi Test Farm at Agricultural Chemical Research Center of Sumitomo Chemical Co.*

Mitsuo Okano, *Professor, Tokyo Women's Medical College*

Masayoshi Watanabe, *Assistant Professor, Yokohama National University Department of Engineering, Division of Material Engineering*

### *Contributors*

Aizo Yamauchi, *President, International Research Exchange Center of Japan Society of Promotion for Industrial Technology*

Yoshihito Osada, *Professor, Department of Scientific Research in Biology at Hokkaido University Graduate School*

Hidetaka Tobita, *Assistant Professor, Department of Engineering, Material Chemistry Division at Fukui University*

Yutaka Tanaka, *Research Associate, Department of Engineering, Material Chemistry Division at Fukui University*

Shunsuke Hirotsu, *Professor, Department of Life Sciences and Engineering, Division of Organism Structures at Tokyo Institute of Technology*

Mitsuhiro Shibayama, *Professor, Department of Textiles, Polymer Molecule Division at Kyoto University of Industrial Art and Textile*

Hidenori Okuzaki, *Assistant, Department of Chemistry and Biology, Division of Biological Engineering at Yamanashi University*

Kanji Kajiwara, *Professor, Department of Technical Art in Material Engineering at Kyoto University of Industrial Art and Textile*

Yukio Naito, *Head of Research, Biological Research Center for Kao*

(the late) Kobayashi Masamichi, *Honorary Professor, Department of Science, Division of Polymer Molecular Research at Osaka University Graduate School*

Hidetoshi Oikawa, *Assistant Professor, Emphasis of Research on Higher Order Structural Controls in Department of Reactive Controls at Reactive Chemistry Research Center at Tohoku University*

Yositsugu Hirokawa, *Technical Councilor, Science and Technology Promotional Office, Hashimoto Phase Separation Structure Project*

Makoto Suzuki, *Professor, Department of Engineering, Division of Metal Engineering at Tohoku University Graduate School*

Ken Nakajima, *Special Research, Division of Basic Science in International Frontier Research System Nano-organic Photonics Material Research Team at Physics and Chemistry Research Center*

Toshio Nishi, *Professor, Department of Engineering Research, Division of Physical Engineering at Tokyo University Graduate School*

Hidemitsu Kuroko, *Assistant Professor, Department of Life Environment, Division of Life Environment at Nara Women's University*

Shukei Yasunaga, *Assistant, Department of Technical Art in Material Engineering at Kyoto University of Industrial Art and Textile*

Mitsue Kobayashi, *Special Researcher, Tokyo Institute of Technology*

Hajime Saito, *Professor, Department of Science, Division of Life Sciences at Himeji Institute of Technology*

Hazime Ichijyo, *Manager of Planning Office, Industrial Engineering Research Center in Department of Industrial Engineering, Agency of Industrial Science and Technology at Ministry of International Trade and Industry*

Masayoshi Watanabe, *Assistant Professor, Yokohama National University Department of Engineering, Division of Material Engineering*

Kunio Nakamura, *Professor, Department of Agriculture, Division of Food Sciences at College of Dairy Agriculture*

Hideo Yamazaki, *Shial, Inc. (Temporarily transferred from Tonen Chemical Co.)*

Koshibe Shigeru, *Shial, Inc. (Temporarily transferred from Tonen Chemical Co.)*

Hirohisa Yoshida, *Assistant, Department of Engineering, Division of Industrial Chemistry at Tokyo Metropolitan University*

Yoshiro Tajitsu, *Professor, Department of Engineering at Yamagata University*

Hotaka Ito, *Instructor, Division of Material Engineering at National Hakodate Technical High School*

Toyoaki Matsuura, *Assistant, Department of Opthamology at Nara Prefectural Medical College*

Yoshihiko Masuda, *Lead Researcher, Third Research Division of Japan Catalytic Polymer Molecule Research Center*

Toshio Yanaki, *Researcher, Shiseido Printed Circuit Board Technology Research Center*

Yuzo Kaneko, *Department of Science, Division of Applied Chemistry at Waseda University*

Kiyotaka Sakai, *Professor, Department of Science, Division of Applied Chemistry at Waseda University*

Teruo Okano, *Professor, Medical Engineering Research Institute at Tokyo Women's Medical College*

Shuji Sakohara, *Professor, Department of Engineering, Chemical Engineering Seminar at Hiroshima University*

Jian-Ping Gong, *Assistant Professor, Department of Scientific Research, Division of Biology at Hokkaido University Graduate School*

Akihiko Kikuchi, *Assistant, Medical Engineering Research Institute at Tokyo Women's Medical College*

Shingo Matukawa, *Assistant, Department of Fisheries, Division of Food Production at Tokyo University of Fisheries*

Kenji Hanabusa, *Assistant Professor, Department of Textiles, Division of Functional Polymer Molecules at Shinshu University*

Ohhoh Shirai, *Professor, Department of Textiles, Division of Functional Polymer Molecules at Shinshu University*

Atushi Suzuki, *Assistant Professor, Department of Engineering Research, Division of Artificial Environment Systems at Yokohama National University Graduate School*

Junji Tanaka, *Department of Camera Products Technology, Division Production Engineering, Process Engineering Group at Optical Equipment Headquarters at Minolta, Inc.*

Eiji Nakanishi, *Assistant Professor, Department of Engineering, Division of Material Engineering at Nagoya Institute of Technology*

Ryoichi Kishi, *Department of Polymer Molecules, Functional Soft Material Group in Material Engineering Technology Research Center in Agency of Industrial Science and Technology at Ministry of International Trade and Industry*

Toshio Kurauchi, *Director, Toyota Central Research Center*

Tohru Shiga, *Head Researcher, LB Department of Toyota Central Research Center*

Keiichi Kaneto, *Professor, Department of Information Technology, Division of Electronic Information Technology at Kyushu Institute of Technology*

Kiyohito Koyama, *Professor, Department of Engineering, Material Engineering Division at Yamagata University*

Yoshinobu Asako, *Lead Researcher, Nippon Shokubai Co. Ltd., Tsukuba Research Center*

Tasuku Saito, *General Manager, Research and Development Headquarters, Development Division No. 2 of Bridgestone, Inc.*

Toshihiro Hirai, *Professor, Department of Textiles, Division of Raw Material Development at Shinshu University*

Keizo Ishii, *Manager, Synthetic Technology Research Center at Japan Paints, Inc.*

Yoshito Ikada, *Professor, Organism Medical Engineering Research Center at Kyoto University*

Lin Wang, *Senior Scientist, P&G Product Development Headquarters*

Rezai E., *P&G Product Development Headquarters*

Fumiaki Matsuzaki, *Group Leader, Department of Polymer Molecule Science Research, Shiseido Printed Circuit Board Technology Research Center*

Jian-Zhang (Kenchu) Yang, *Researcher, Beauty Care Product Division of P&G Product Development Headquarters*

Chun Lou Xiao, *Section Leader, Beauty Care Product Division of P&G Product Development Headquarters*

Yasunari Nakama, *Councilor, Shiseido Printed Circuit Board Technology Research Center*

Keisuke Sakuda, *Assistant Director, Fragrance Development Research Center at Ogawa Perfumes, Co.*

Akio Usui, *Thermofilm, Co.*

Mitsuharu Tominaga, *Executive Director, Fuji Light Technology, Inc.*

Takashi Naoi, *Head Researcher, Ashikaga Research Center of Fuji Film, Inc.*

Makoto Ichikawa, *Lion, Corp. Better Living Research Center*

Takamitsu Tamura, *Lion, Corp. Material Engineering Center*

Takao Fushimi, *Examiner, Patent Office Third Examination Office at Ministry of International Trade and Industry*

Kohichi Nakazato, *Integrated Culture Research Institute, Division of Life Environment (Chemistry) at Tokyo University Graduate School*

Masayuki Yamato, *Researcher, Doctor at Japan Society for the Promotion of Science, and Japan Medical Engineering Research Institute of Tokyo Women's Medical College*

Toshihiko Hayasi, *Professor, Integrated Culture Research Institute, Division of Life Environment (Chemistry) at Tokyo University Graduate School*

Naoki Negishi, *Assistant Professor, Department of Cosmetic Surgery at Tokyo Women's Medical College*

Mikihiro Nozaki, *Professor, Department of Cosmetic Surgery at Tokyo Women's Medical College*

Yoshiharu Machida, *Professor, Department of Medical Pharmacology Research at Hoshi College of Pharmacy*

Naoki Nagai, *Professor, Department of Pharmacology at Hoshi College of Pharmacy*

Kenji Sugibayashi, *Assistant Professor, Department of Pharmacology at Josai University*

Yohken Morimoto, *Department Chair Professor, Department of Pharmacology at Josai University*

Toshio Inaki, *Manager, Division of Formulation Research in Fuji Research Center of Kyowa, Inc.*

Seiichi Aiba, *Manager, Department of Organic Functional Materials, Division of Functional Polymer Molecule Research, Osaka Industrial Engineering Research Center of Agency of Industrial Science and Technology at Ministry of International Trade and Industry*

Masakatsu Yonese, *Professor, Department of Pharmacology, Division of Pharmacology Materials at Nagoya City University*

Etsuo Kokufuta, *Professor, Department of Applied Biology at Tsukuba University*

Hiroo Iwata, *Assistant Professor, Organism Medical Engineering Research Center at Kyoto University*

Seigo Ouchi, *Head Researcher, Agricultural Chemical Research Center at Sumitomo Chemical Engineering, Co.*

Ryoichi Oshiumi, *Former Engineering Manager, Nippon Shokubai Co. Ltd. Water-absorbent Resin Engineering Research Association*

Tatsuro Toyoda, *Nishikawa Rubber Engineering, Inc. Industrial Material Division*

Nobuyuki Harada, *Researcher, Third Research Division of Japan Catalytic Polymer Molecule Research Center*

Osamu Tanaka, *Engineering Manager, Ask Techno Construction, Inc.*

Mitsuharu Ohsawa, *Group Leader, Fire Resistance Systems Group of Kenzai Techno Research Center*

Takeshi Kawachi, *Office Manager, Chemical Research Division of Ohbayashi Engineering Research Center, Inc.*

Hiroaki Takayanagi, *Head Researcher, Functional Chemistry Research Center in Yokohama Research Center of Mitsubishi Chemical, Inc.*

Yuichi Mori, *Guest Professor, Department of Science and Engineering Research Center at Waseda University*

Tomoki Gomi, *Assistant Lead Researcher, Third Research Division of Japan Catalytic Polymer Molecule Research Center*

Katsumi Kuboshima, *President, Kuboshima Engineering Company*

Hiroyuki Kakiuchi, *Mitsubishi Chemical, Inc., Tsukuba Research Center*

Baba Yoshinobu, *Professor, Department of Pharmacology, Division of Pharmacological Sciences and Chemistry at Tokushima University*

Toshiyuki Osawa, *Acting Manager, Engineer, Thermal Division NA-PT at Shotsu Office of Ricoh, Inc.*

Kazuo Okuyama, *Assistant Councilor, Membrane Research Laboratory, Asahi Chemical Industry Co., Ltd.*

Takahiro Saito, *Yokohama National University Graduate School, Department of Engineering, Division of Engineering Research*

Yoshiro Sakai, *Professor, Department of Engineering, Division of Applied Chemistry at Ehime University*

Seisuke Tomita, *Managing Director, Development and Production Headquarters at Bridgestone Sports, Inc.*

Hiroshi Kasahara, *Taikisha, Inc. Environment System Office*

Shigeru Sato, *Head Researcher, Engineering Development Center at Kurita Engineering, Inc.*

Okihiko Hirasa, *Professor, Iwate University*

Seiro Nishio, *Former Member of Disposable Diaper Technology and Environment Group of Japan Sanitary Material Engineering Association*

# VOLUME IV

## Environment:
## Earth Environment & Gels

# CHAPTER 1

# Environmental Preservation

*SHIGERU SATO*

## Chapter contents

# 1 INTRODUCTION

In recent years, global warming, ozone layer destruction, acid rain, deforestation, and soil erosion have become serious issues. As a result, the world's interest in preservation and environmental protection of the Earth has increased. In Japan, air pollution, a paucity of garbage dump sites due to land shortages, environmental pollution due to illegal dumping of garbage, river and lake pollution, and declining water quality as a result of fertilizer runoff are problems closely related to daily life. Environmental protection is an important subject that requires immediate attention if humankind is to continue and flourish.

Polymeric materials (called polymeric flocking agents), which function by gelation in a broad sense because they are not actually gels, are used in preserving the environment. These polymeric flocking agents are water-soluble polymers used to clean water and treat waste. They are indispensable in our society for lake clean-up, as well as of industrial waste, sewage disposal, and in other areas.

Polymeric flocking agents will be reviewed here with respect to types and structures, functional mechanisms, and actual application examples.

**Table 1** Relationship between particle size and rate of sedimentation [3].

Temperature: 20 °C

| Particle size (mm) | | Density 2.00 | | Density 1.02 | |
|---|---|---|---|---|---|
| | | Rate of sedimentation | | Rate of sedimentation | |
| | | cm/s | Time necesssary for 1 m of sediment | cm/s | Time necesssary for 1 m of sediment |
| 5 | Aggregated flock | — | — | 20.8 | 4.8 s |
| 1 | Aggregated flock, sand | 42 | 2.4 s | $8.4 \times 10^{-1}$ | 2.0 min |
| 0.1 | Silt, clay | $4.2 \times 10^{-1}$ | 4.0 min | $8.4 \times 10^{-3}$ | 3.3 h |
| 0.01 | Bacteria | $4.2 \times 10^{-3}$ | 6.6 h | $8.4 \times 10^{-5}$ | 13.8 D |
| 0.001 | Colloid | $4.2 \times 10^{-5}$ | 28 D | $8.4 \times 10^{-7}$ | (3.8 yr) |
| | | $4.2 \times 10^{-7}$ | (7.5 yr) | $8.4 \times 10^{-9}$ | (380 yr) |

Stoke's equation

$$V = \frac{g_n \cdot D_p^2 (\rho_p - \rho)}{18\eta}$$

$V$: Sedimentation rate
$g_n$: Acceleration of gravity
$D_p$: Particle diameter
$(\rho_p - \rho)$: Density differences of particles and water
$\eta$: Water viscosity

**Table 2** The size of suspension particles and applicable flocking agent range.

| Particle size | $10^{-1}$ cm<br>**1 mm** | $10^{-2}$<br>**100** $\mu$ | $10^{-3}$<br>**10** $\mu$ | $10^{-4}$<br>**1** $\mu$ | $10^{-5}$<br>**100m** $\mu$ | $10^{-6}$<br>**10m** $\mu$ | $10^{-7}$<br>**1m** $\mu$ | $10^{-8}$ 1Å |
|---|---|---|---|---|---|---|---|---|
| Type of particles | Large-particle sand, small-particle silt<br>Fiber waste | | | | Colloid<br>Bentonite | Dyes | | |
| | | | | Bacterica | | | | |
| | | | | | Virus<br>Proteins | | Low molecular weight compounds<br>and ions | |
| | | | | | | | Surface active agents | |
| | | Starch | | | | | | |
| | | | Cosmetic powder | | | | | |
| Range | | | Single use of polymeric flocking agent | | | | | |
| | | Combined use of an inorganic and a polymeric flocking agent | | | | | | |

## 2   WHAT IS A POLYMERIC FLOCKING AGENT?

A polymeric flocking agent is a polyelectrolyte used both to aggregate microparticles that are suspended in water and increase particle size [1, 2]. Polymeric flocking agents facilitate the separation of liquids from solids in a suspension by using flocking behavior. Used to clean both water and wastewater, their main use is in wastewater treatment processes. As shown in Table 1 [3], particle size influences significantly the rate of sedimentation (Stokes' equation). An increase in particle size makes liquid-solid separation easier. Table 2 lists the particle size ranges for which polymeric flocking agents can be effective [2].

In the history of polymeric flocking agents, starch and gelatin were used as flocking agents in the 1920s. In 1954, both Dow Chemical and American Cyanamide began to manufacture poly(acrylamide) (the first synthetic polymeric flocking agent). Japan started using this material when a domestic manufacturer began production, with tremendous increases in the use of polymeric flocking agents after pollution problems became serious. This was particularly true after a water pollution protection law was enacted in the 1970s.

Today, production is as high as 29,000 metric tons. For reference, Fig. 1 shows the trend in domestic use of polymeric flocking agents [4].

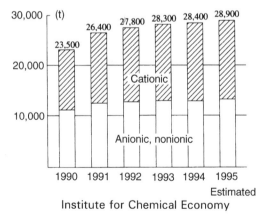

**Fig. 1**  Domestic consumption of polymeric flocking agents [4].

## 3   TYPE AND STRUCTURE OF POLYMERIC FLOCKING AGENTS [3]

Polymeric flocking agents can be classified as nonionic, anionic, cationic, and amphoteric.

### 3.1   Nonionic and Anionic Polymers

The chemical structures of the nonionic and anionic polymers are shown in Fig. 2. The main chemical structure is poly(acrylamide). Depending on the characteristics of the wastewater and treatment conditions, hydrolysis or copolymerization with acrylic acid is carried out to adjust anion concentrations (0–100 mol%). In addition to carboxylic acid a sulfonic type is also used.

The main monomer, acrylamide, readily polymerizes and yields a linear polymer. In general, the polymers used have ultrahigh molecular weights ranging from 10–20 million. Aqueous solutions in the range 20–50wt% monomer range are usually used for polymerization. Upon polymerization, the polymers are dried to powders. Some manufacturers

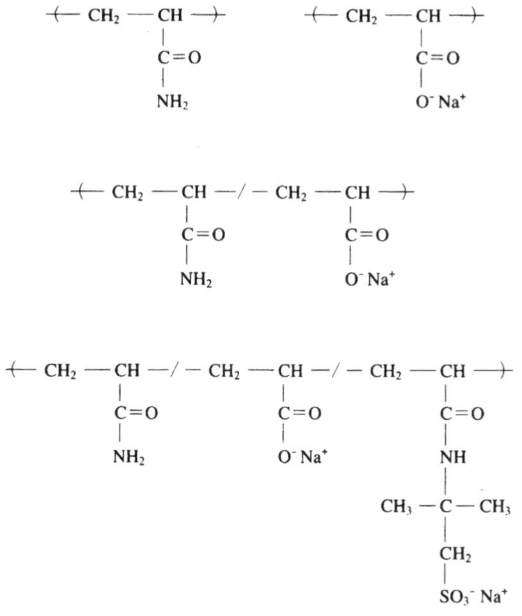

**Fig. 2**   Chemical structure of nonionic and anionic polymeric flocking agents.

produce these polymers by reverse-phase suspension polymerization and others use reverse-phase emulsion polymerization.

## 3.2 Cationic Polymers

The discussion of cationic polymers here is based on the polymerization reaction type.

### 3.2.1 Polymerization type

Representative cationic polymeric flocking agents are shown in Fig. 3. Acrylic acid and an aminoalkylester of methacrylic acid are widely used cationic monomers. In addition, tertiary amine salts and quaternary ammonium salts like ethylchloride and benzylchloride are used. As the concentration of cations needs to be changed depending on the characteristics of the wastewater or treatment conditions, copolymers that fall between these cationic monomers and the acrylamide ones are also used. These polymers are manufactured by polymerization of aqueous solutions, reverse-phase suspension polymerization, and reverse-phase emulsion polymerization. Commercial products are available as powders, aqueous solutions, and emulsions. Traditionally, the molecular weight was limited to approximately 10 million. However, due to advance in polymerization techniques, higher molecular weight polymers have recently

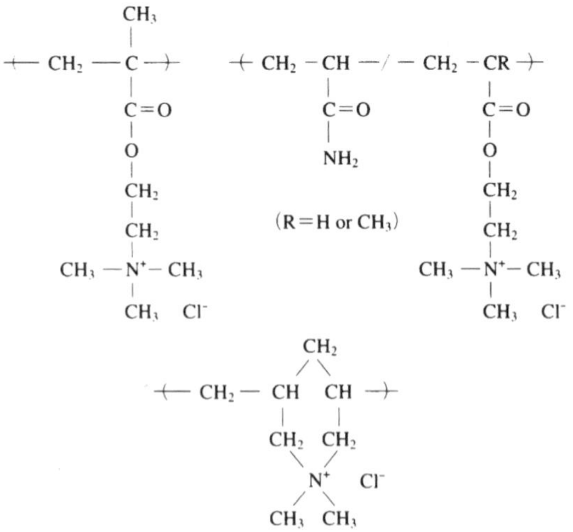

**Fig. 3** Chemical structure of representative cationic polymeric flocking agents.

become available. Another monomer, a difunctional diacryldimethylam-monium chloride, which forms a quaternary ammonium-type cationic polymer via ring opening polymerization, can also be used. However, due to the chain transfer reaction of the aryl group, it is difficult to obtain high molecular weight polymers. Thus, practical application examples are limited.

### 3.2.2   Modified polymers

The chemical structures of representative modified cationic polymers are shown in Fig. 4. Representative polymers are aminomethylated acrylamide homopolymer (Mannich modification) with formaldehyde and dimethyl-amine, and Hoffman-decomposed poly(acrylamide) made from the reaction of an acrylamide homopolymer with sodium perchlorate at relatively low temperature. Because the former polymer (Mannich-modified type) does not show molecular weight reduction during modification, high molecular weight modified polymers result. However, the latter example (Hoffman-decomposed type) has a shortcoming: molecular weight is

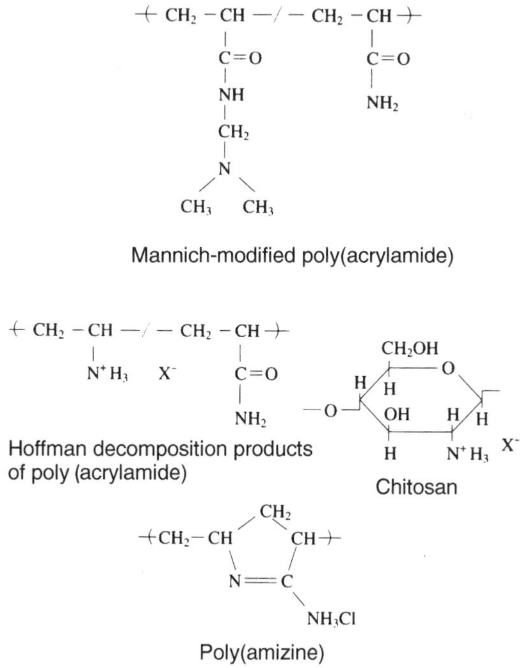

Fig. 4   The chemical structure of representative modified cationic polymers.

reduced due to the use of sodium perchlorate as a halogenated agent during the Hoffman reaction. Because these two polymers are modified under basic conditions, the amide groups may be partially hydrolyzed to form carboxylic groups. In this case, an amphoteric polymer can be made.

Polyamizine, which can be obtained by modifying the copolymer of N-vinyl formamide and acrylonitrile by an acid or alkali, has also been used in recent years [5]. This polymer has high molecular weight and high concentration of cations. A natural polymer, chitosan is used as a cationic polymer. Chitosan is obtained by acetylating chitin, which is contained in the shells of crabs and shrimps.

### 3.2.3 Condensation polymers

Poly(ethylene imide), a condensation product of epichlorohydrin and dimethylamine, is a representative polymer in this class whose chemical structure is shown in Fig. 5. It is extremely difficult to obtain molecular weights similar to those of the polymers shown in Section 3.2.1. Ordinarily, the molecular weights of these types of polymers are several tens of thousands to several hundreds of thousands.

Although these condensation polymers are in a broad sense flocking agents, the ability to aggregate is weak due to their low molecular weights.

Characteristic of these polymers is that cationic concentration per weight is high. These compounds are used as coflocking agents to neutralize the electrical charges of suspended particles in a manner similar to an inorganic flocking agent such as aluminum sulfate or poly(aluminum chloride). They are also called "organic aggregating agents."

### 3.3 Amphoteric Polymers

Among the amphoteric polymers, there is a copolymer type in which cationic and anionic groups coexist in the same polymer chain, and a polymer blend type in which a cationic polymer and an anionic polymer are blended.

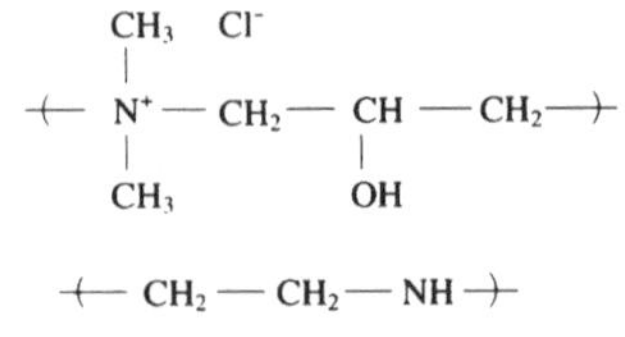

$$
\begin{array}{c}
CH_3 \quad Cl^- \\
| \\
\text{---}\ N^+ \text{---} CH_2 \text{---} CH \text{---} CH_2\text{---} \\
| \qquad\qquad | \\
CH_3 \qquad\quad OH
\end{array}
$$

$$
\text{---}\ CH_2 \text{---} CH_2 \text{---} NH \text{---}
$$

**Fig. 5** The chemical structure of representative condensation cationic polymers.

### 3.3.1   Copolymer type

The aminoalkylester of (metha)acrylic acid discussed in Section 3.2.1 is a representative cationic monomer. For the anionic monomer, acrylic acid is used, and for the nonionic monomer, acrylamide is used. Concentrations of cations, anions, and the cation/anion ratio are adjusted by controlling their monomer ratio and molecular weight in relation to the wastewater, other waste characteristics, and the treatment device that is used.

### 3.3.2   Blend type

A blend is made by a cationic polymer (shown in Section 3.2.1) and an anionic polymer (shown in Section 3.1). It will be necessary to adjust pH in order to prevent formation of an insoluble complex from the reaction of the cations and anions.

## 4   FUNCTIONAL MECHANISMS (FLOCKING MECHANISMS)

The conformation of polymer flocking agent molecules dissolved in water is strongly influenced by the conditions of the solution [1, 2]. Polymer molecules tend to spread by electrical repulsion of the dissociated groups. The spread and conformation depend strongly on both the concentration of salt that coexists in the solution and the pH. As the salt concentration increases, the molecule shrinks. This is due to the shielding of the electric charge and, consequently, the suppression of the repulsive force among dissociated groups. Further, the pH changes the degree of dissociation. Hence, the conformation of the polymer chain also changes as the electric repulsive forces among dissociated groups change.

If suspended particles with an opposite electrical charge from that of the dissociated groups of the polymer in water, the polymer molecules adsorb onto the particle surface by electrical attractive forces while bridging the particles. Upon adsorption, the electrical charges of the adsorbed molecules are neutralized and the polymer shrinks due to the reduction in electrical repulsive forces, which causes aggregation of the suspended particles (flocking). This flocking is an irreversible inter- and intramolecular crosslinking reaction through the suspended particles. In a broad sense, this phenomenon can be viewed as "gelation." Due to the variation of application methods based on the ionic nature of the polymer, the functional mechanisms will be described with respect to the ionic nature of the polymers.

### 4.1   Nonionic and Anionic Polymers

In general, organic waste and waste sludge suspension are negatively charged and, thus, nonionic and anionic polymers cannot adsorb onto the particles. Therefore, if a nonionic or anionic polymer is used, an inorganic flocking agent such as aluminum sulfate, poly(aluminum chloride) or iron chloride (II) must first be added to the wastewater to cause microflock by the metal hydroxides. At this time, the microflock surface is slightly positively charged by the metallic ions on which the polymer adsorbs via ionic bonding, thereby resulting in flocking (see Fig. 6). The molecular weight of nonionic and anionic polymers is extremely large (10,000,000–20,000,000) and the molecule takes widely spread conformations. Hence, the ability to flock suspended particles is high. However, the mechanical strength of the formed flock is generally weak. Consequently, nonionic and anionic polymers are used to clean wastewater suspensions and purify water.

Flocking takes place by the adsorption of a polymer via ionic or hydrogen bonding among the carboxylic or amide groups of the polymer and the suspended particle surface. This results in the subsequent bridging of particles. There are several studies on the adsorption of polymers on the suspended particles [6–8]. Polymers exist as random coils in water. However, upon adsorption onto the particles, part of the polymer chain

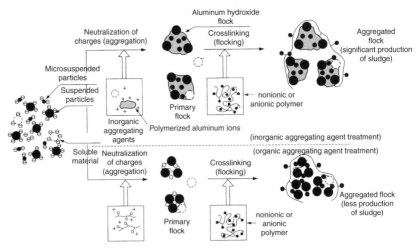

**Fig. 6**   Flocking by nonionic and anionic polymers [3].

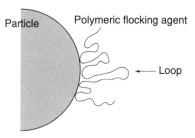

**Fig. 7** A model of adsorbed polymer [3].

adsorbs onto the particle surface while the remainder of the chain extends into water as a loop (see Fig. 7).

As a model for flocking mechanism, the Michaels' [9] "adsorption–bridging" concept as shown in Fig. 8 is generally accepted.

The appropriate concentration of anions on the polymer depends on the flocking conditions of wastewater (such as pH) (Fig. 9). This is because the state of dissociation on the polymer chain changes. The effect of polymer molecular weight is also shown in Fig. 10. In general, the greater the molecular weight, the wider the spread of the polymer, which will thus exhibit better flocking ability. However, polymer dispersion sometimes becomes problematic if the molecular weight is too high. The condition of the wastewater being treated plays an interactive role in this problem. Generally speaking, if the molecular weight distribution is narrow, the polymer exhibits better flocking ability. The effect of the

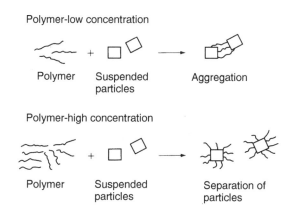

**Fig. 8** "Adsorption-Bridging" model (Michaels) [9].

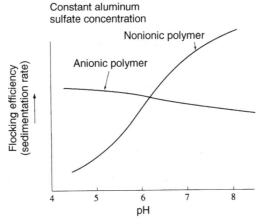

**Fig. 9**   The effect of treatment pH and polymer flocking agent.

polymer branches has not been studied extensively. However, polymer spread is greater for more linear polymers and shows better flocking ability.

## 4.2   Cationic Polymers

Cationic polymers directly adsorb ionically onto the negatively charged suspended particles in organic waste and soil waste and cause aggregation [3]. Cationic polymers can adsorb tightly to particles in suspension. As the

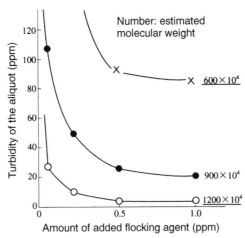

**Fig. 10**   Molecular weight of poly(acryl amide) and the turbidity of the aliquot of flocked wastewater.

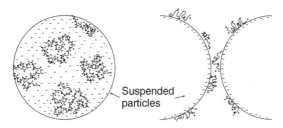

**Fig. 11** The flocking model proposed by Gregory [10].

mechanical strength of the flocks is greater than those prepared from nonionic or anionic polymers, they are used during dehydration of biotreated wastewater (mechanical strength of flocks is required), human waste, and general factory wastewater. Therefore, cationic polymers are often called "dehydrating agents for wastes."

The aforementioned Michaels' [9] "adsorption-bridging" concept and Gregory's [10] "mosaic electrical charge model" (Fig. 11) are proposed as flocking mechanisms. However, these mechanisms sometimes do not describe the real situation. For example, Gregory's concept [10] cannot explain the fact that the larger the molecular weight the greater the ability to form flock or that the high concentration of cations does not necessarily lead to larger flocks.

Accordingly, it is proposed that flocking by cationic polymers is not only influenced by the suspended particles but also by the anionic polymeric materials (polysaccharides or proteins generated by microbes) dissolved in the waste [3]. This anionic polymer acts as a binder (it adsorbs ionically with the flocking agent and strengthens the interparticle bridging) and flocking of the suspended particles takes place (see Fig. 12).

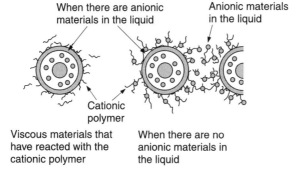

**Fig. 12** The flocking model for cationic polymers [3].

As the flocked suspended particles of the waste are mechanically dehydrated by a dehydration device, the generated flocks are required to possess the mechanical strength to withstand the mechanical forces (pressure and centrifugal forces). In order to generate mechanically strong flocks, it is necessary to increase the molecular weight of the polymer. However, if the molecular weight is too high, the suspension of the polymer in the waste becomes poor and can lead to lower flocking ability. Hence, it is important to select an appropriate molecular weight suitable for the type and structure of the dehydration device. In general, the narrower the molecular weight distribution the better the adsorption efficiency and, thus, the greater the flocking strength.

### 4.3  Amphoteric Polymers

Amphoteric polymers are used for the dehydration process as in the case of cationic polymers. As was already discussed in Section 4.2, there are two cases in which only the polymer is used as is or an inorganic flocking agent is added first, followed by the addition of an amphoteric polymer used as the secondary treatment agent. The method used is determined by the characteristics of the waste or the type and capacity of the treatment device (dehydration device).

The flocking mechanism of amphoteric polymers is similar to the cationic polymers when they are used alone. As there are anionic portions in the polymer chain, there will be strong intermolecular ionic interactions among polymer chains, resulting in even stronger flocks. When they are used along with an inorganic flocking agent, the inorganic agent first adsorbs onto the suspended particle surfaces and anionic polymeric materials in the liquid. Thus, both cationic and anionic groups can adsorb onto the particles. Furthermore, due to the intermolecular ionic bonding, an irreversible polymer complex is formed and a stronger flock can also be formed.

## 5  ACTUAL APPLICATION EXAMPLES

### 5.1  Flocking Treatment of Wastewater

The main reason for the use of nonionic and anionic polymers is to clean water. For nonionic and anionic polymers, the treatment flow diagram for the total wastewater of paper pulp illustrated in Fig. 13 provides us with an application example. As there are dissolved organic materials in the

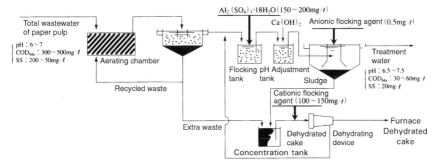

**Fig. 13** Total wastewater treatment flowchart of paper pulp.

wastewater along with the suspended particles, the waste is biologically treated in the aerating chamber and biochemical oxygen demand (BOD) components are decomposed. After this treatment, the suspended particles are sedimented, separated, and treated as extra waste. However, as there are still fine suspended particles remaining in the aliquot, they are cleaned by the polymeric flocking agent.

Following biological treatment and after suspended particles have been sedimented, an inorganic flocking agent (aluminum sulfate, Fig. 13) is added to the aliquot. Upon adjusting the pH to neutral with sodium hydroxide, an anionic polymer will be added. Fine suspended particles will be agitated slowly in a thickener and sedimented. The separated, sedimented material is called aggregated sedimentation waste. This material is sent through the waste treatment process along with the excess waste already described. The actual waste treatment will be described in the following section.

## 5.2 Waste Treatment [11]

In order to decompose the soluble organic components in wastewater, biological treatment is generally applied to sewage, ordinary factory wastewater, and human waste. The biological treatment uses aerobic microbes (the active waste sludge method) or digestion of organic materials by anaerobic microbes. After the biological treatment, waste sludge will always be generated. It consists of over 99% water and suspended particles, with a solid content of $< 1\%$. The suspended particles are the microbes themselves and their dead bodies, and natural materials such as proteins and polysaccharides that were generated during the growth process of the microbes. The suspended particles are negatively charged hydrophilic colloids and are extremely difficult to sediment due to the electrical repulsion between particles. As the solid waste is eventually thrown out, its volume should be reduced as much as possible. Thus, it is essential to remove water from the solid waste. In general, cationic polymers are used mainly to remove water (dehydrate) from solid waste.

### 5.2.1 Dehydration of solid waste by only a cationic polymer

As an example of the dehydration of solid waste, the sewage flow chart is shown in Fig. 14. The sewage from individual homes and factories is first treated to remove the large suspended particles in either the sand pond or the first sedimentation pond. Then biological treatment follows. The solid wastes generated during this process (extra waste and biological waste) are mixed and compressed by gravity until the solid concentration becomes 1–2%. Added to this concentrated waste is a cationic polymer of 0.5–1.0%, basing the percentage that is used on the solid waste component to flock; dehydration follows. This solid waste is called dehydrated cake (solid content of 20–30%). Eventually, this material is burned or disposed of at a landfill.

In order to remove moisture effectively from viscous materials such as protein or acidic polysaccharides, which contain large amounts of water, it is necessary to: (1) neutralize the negative charges on the suspended particles (neutralize the negative charges of the viscous anionic polymers in order to shrink the polymers and help water to escape); and (2) flock the suspended particles (help solid–liquid separation by increasing particle size).

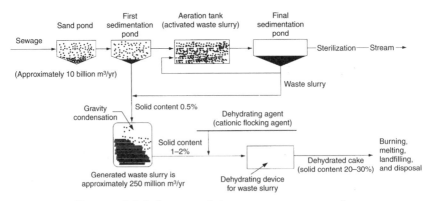

**Fig. 14**   Brief diagram of the sewage treatment flow.

The higher concentration of the cations exhibits better neutralization of the negative charges of the viscous material, leading in turn to more efficient reduction in water content in the waste slurry. In addition, higher molecular weight shows a stronger ability to flock and, thus, solid–liquid separation is easier.

### 5.2.2 Dehydration of waste slurry by combined use of cationic and anionic polymers [12]

As previously described, the dehydration of waste slurry requires: (1) neutralization of the negative charges on the suspended particles; and (2) flocking of the suspended particles. In simple treatment with only a cationic polymer, the polymer can perform these two functions. However, the use of both cationic and anionic polymers to dehydrate waste slurry is accomplished by having each one perform one function. Neutralization is achieved by the cationic polymer and the suspended particles are flocked by the anionic polymer. Figure 15 illustrates the schematic diagram for this mechanism.

### 5.2.3 Dehydration of waste slurry by the combined use of an inorganic flocking agent and an amphoteric polymer

Basically, the approach is similar to the way cationic and anionic polymers are used as described in Section 5.2.2. Neutralization is achieved by an inorganic flocking agent and the flocking of the suspended particles is achieved by an amphoteric polymer. If the ability to neutralize charges is compared between the inorganic and cationic polymers, the former has a

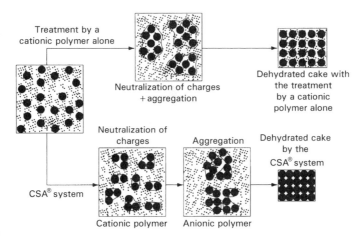

**Fig. 15**

stronger ability to neutralize. This is because the inorganic flocking agent contains multivalent metals (aluminum, iron, etc.) and is smaller in size than the polymer. Thus, the technique described here is the most efficient dehydrating method. The functional mechanism has already been described.

## 6  CONCLUSIONS

Among various environmental issues, the most important subject for the survival of mankind is the preservation of water resources, effective use of water, and wide spread installation of sewerage in Japan (at the present it is about 50%). There is no doubt that the role of polymeric flocking agent will be more important in the future. Hence, related companies are currently seeking materials and conducting developmental research for improved performance of these polymers. We are not generally familiar with polymeric flocking agents. However, as described thus far, they are important functional polymers for preservation of our environment. The author is happy if the readers develop further interest and understanding of polymeric flocking agent through this article.

## REFERENCES

1   Oomori, E. (1973). *Polymeric Flocking Agents*, Kobunshi Kanko-kai.
2   Nagasawa, M. and Takizawa, A. (1976) *Polymeric Water Treatment Agents*, Chijin Shoin.
3   Kogyo, K. (1995). *Handbook on Drugs*, 3rd ed.
4   (1995). *Chemical Economics*, Special Issue, Institute for Chemical Economics.
5   (1994). For example Mitsubishi Chemicals, Tokkyo Kaiho Heisei 6-218400.
6   Takahashi, A. (1985). *Hyomen* **23**: 158.
7   Ruehrwein, R.A. and Ward, D.W. (1952). *Soil Sci.* **73**: 485.
8   Healy, T.W. and LaMer, V.K. (1964). *J. Colloid Sci.* **19**: 323.
9   Michaels, A.S. (1954). *Ind. Eng. Chem.* **46**(7): 1485.
10  Gregory, J. (1973). *J. Colloid Interface Sci.* **42**(2): 448.
11  Hashimoto, M. and Hiraoka, M. (1990). *Wat. Sci. Tech.* **22**(12): 143.
12  Wakeuchi, T. and Hashimoto, M. (1985). *Preprint of IAWPRC's First Asian Conference*, p. 448.

# CHAPTER 2

# *Forestation Technology*

*IOKIIHIKO HIRASA*

## Chapter contents

# 1  INTRODUCTION

The large-scale fires in the Kuwait oilfields during the Persian Gulf War and the Pinatubo volcanic eruptions have been the cause of much recent concern. While pollution during the 1970s was on a regional scale, pollution since the late 1980s is characterized by its wide scale nature, crossing borders to destroy the ozone layers, and to create global warming and acid rain. Deforestation, one of these environmental problems, is not recognized as a daily problem in Japan due to its monsoon climate. However, as it is one aspect of world pollution, we cannot be indifferent to global environmental disruptions. Further, as one of the main economic world powers, we must admit that Japan is contributing to this trend.

There have been many technological advances in Japan to cope with the environmental pollution of the 1970s. These advances should be used to help other countries face their current problems.

The gradual move from green environments to deserts as a result of humanity-caused environmental dislocations is not a new phenomenon. Many old archeological sites have been unearthed in the Sahara desert, all over the Middle East, and in Takramakarn. Desert areas have increased in just 1000 years. However, the deforestation that has been of recent concern is due to increased meadowlands, sudden increases in population, reductions of forests to obtain fuel and fields, and abandonment of soil maintenance as a result of local conflicts. With such multifaceted problems, to what extent can science and technology, in particular polymer science, contribute to reforestation? It is perhaps difficult for reforestation to succeed once the area is completely deforested. However, it is possible to produce vegetables in desert areas as well as to reproduce plants that help prevent deforestation.

# 2  DESERT AND CLIMATE

The desert is not necessarily made only of sand. Sometimes it consists of rocks, gravel, and soil of finer consistency than sand. Thus deserts are classified into rock deserts, sand deserts, gravel deserts, and soil deserts, depending upon composition. In the low- to mid-latitudes, temperature is high and temperature variation in a year or in a day is quite large. In fact, a daily temperature difference can be on the order of 40°C.

To express dryness in a certain region, Thornthwaite's humidity index (Im) is used and is expressed as follows [1]:

$$\text{Humidity index (Im)} = (100s - 60d)/n$$

s: annual excess water
d: annual deficient water
n: annual evaporation of water

The annual evaporation of water means total possible evaporation of water, including evaporation from both the surface and the plants.

Meigs classified Im of from $-40$ to $-60$, where there is no precipitation over 12 mo, as extreme dryness; an Im of from $-20$ to $-40$ is considered half-dryness. The rest is considered to be "dryness." Over the entire globe, 4.3% of the land is in the extreme-dryness category, 16.2% is in the dryness category, and 14.6% is in the half-dryness category. Hence, 30% of all land has dry conditions. The annual precipitation measure used to distinguish desert from a half-dry region is 250 mm.

According to the definition from the United Nations Deforestation Conference held in 1977, deforestation is the state in which the destruction of ecological systems of dry and half-dry lands due to development, efforts to increase food production, and the introduction of new technology to support population increases, resulting in ongoing increases in desert environments [2].

Aside from extreme-dryness deserts like the Sahara and others, deforestation is spreading throughout dry and half-dry lands [1]. In these dry and half-dry lands, there are many places where annual precipitation is only 200–300 mm. Due to daily temperature variations, the temperature may drop below 0°C and fog may be generated. The fog may make it possible to grow plants if this condition is utilized effectively.

## 3 IRRIGATION FOR GROWING PLANTS

Aside from poor-quality soil, deserts lack water to grow plants. Hence, a dam or underground water irrigation is needed. Evaporation of water in the desert causes the soil to be salt-laden due to the capillary effect.

If water penetration of soil is within normal boundaries, the critical salt concentration of the water used for irrigation will be < 1000 ppm. In areas where salt concentration of both soil and irrigation water is high, effective irrigation methods include surface, sprinkler, and drip irrigation.

Surface and sprinkler irrigation are used for relatively water-rich regions and drip irrigation is used for the water-poor regions.

## 4   SUPERABSORBENT POLYMERS (SAP) AS SOIL MODIFIERS

At the forest surface where fallen leaves have accumulated over time and porous peat has formed due to microbial action, the rate of penetration is 400 mm/h, resulting in high-water retention. Exposed land by contrast will have a rate of 80 mm/h [3]. The fallen leaves are digested by microbes and turn into partially hydrophilic polymeric colloids. Plant fibers and inorganic materials including clay are added to this natural mixture. The result is a soil that has good aeration, permeation, and water retention.

However, it is impossible to expect formation of soil rich in organic matter in dry areas because it is difficult for poor soil to hold water. Furthermore, irrigated water evaporates far more quickly than water that evaporates from plants. Hence, the water is not effectively used. To improve water retention over time by suppressing direct evaporation and soil permeation, superabsorbent polymers have been proposed [4].

Superabsorbent polymers (SAP) were first synthesized in 1975 at the Institute of the U.S. Department of Agriculture from graft polymers made of cornstarch and acrylonitrile. Since then, superabsorbent synthetic polymers made of cellulose, polyacrylic acid, and poly(vinyl alcohol)-type have been developed. Currently, these polymers are used in large quantities for sanitation purposes.

Superabsorbent polymers or superabsorbent polymer gels are mostly crosslinked polymers that have a high concentration of anions such as carboxylic acid. These ionic groups are hydrated and a large amount of water is absorbed. The polymer then swells, forming a hydrogel.

The degree of swelling of superabsorbent polymers can be expressed by the following equation using Flory's swelling theory of ionic networks [5]:

$$Q_m^{5/3} = v \cdot M_c \left( 1 - 2\frac{M_c}{M} \right) \left[ \frac{(i/2V_u)^2}{S} + \frac{(1/2 - \chi_1)}{V_1} \right]$$

$$= \frac{\left[ \dfrac{(i/2V_u)^2}{S} + \dfrac{(1/2 - \chi_1)}{V_1} \right]}{v_e/V_0}$$

where $v$ is the specific volume of the polymer, $S$ is the ionic strength of the external solution, $M$ is the number average molecular weight of the polymer, $M_c$ is the molecular weight between the crosslink points, $i/V$ is the ionic charge density fixed on the polymer networks, $\chi_1$ is the interaction parameter between the polymer and the solvent, $V_1$ is the molar volume of the solvent, $V_0$ is actual volume of the polymer, and $v_e$ is the number of the effective chain in the networks.

The first term on the right-hand side of this equation depends on the osmotic pressure of the ions, and the second term on the interaction between the polymer and water, and $v_e/V_0$ on the crosslink density. Hence, the degree of swelling increases as the electron charge density of the networks increases but decreases as the ionic strength of the external solution and the crosslink density both increase. Masuda *et al.* reported on the water uptake of starch/acrylic acid copolymer (superabsorbent polymers) and the effect of salt concentration and crosslink density on water uptake [6]. They found that the polymer exhibited high water uptake of 300–1000 g-$H_2O$/g-polymer. The bound water was 1.18 g and half-bound water was 0.38 g. Accordingly, the majority of the absorbed water was free water. When superabsorbent polymers are used as a soil modifier, this free water contributes greatly to their usefulness. In order to support such hypotheses, Toyama and other researchers as well, have been active in the application of superabsorbent polymers for low water use forms of irrigation farming in both deserts and dry lands [4, 7].

Soil science wisdom holds that plant growth is influenced by the distribution relationship of soil (solid phase), water (liquid phase), and air (gas phase) in the soil bed. In a properly composed soil bed, inorganic soil and organic polymers (as a result of plant digestion) form aggregates. These aggregates are porous and have the ideal 3-phase distribution of solids, liquid, and gas. On the other hand, if there is little digested plant matter and water uptake of the soil is poor, plant growth will suffer.

To determine how to supplement low water uptake, Toyama studied the water content, 3-phase distribution, and the relationship between the amount of superabsorbent polymer and the growth of plants [8].

Although the addition of superabsorbent polymers dramatically increases water uptake, it is achieved by sacrificing the gas phase. Thus, too much superabsorbent polymer has a negative effect. To illustrate—an optimal concentration of polymer was used in the cultivation of spinach.

In dry areas, the vegetable harvest increased by 1.5 times when 0.1% of the polymer was mixed into the soil [4].

As part of the green earth project, The Japanese Society for Desert Utilization together with Egypt are planning reforestation of desert land using a synthetic soil composite made of a superabsorbent polymer and bentnite. This project has attracted interest as to whether plant growth can be manipulated by microenvironmental processes. However, some problems arise when such a polymer is used in large quantities, including biodegradability, transfer, and accumulation of the polymer.

## 5  RECIRCULATION OF WATER THROUGH PLANTS

We have discussed thus far how evaporation of water supplied by irrigation can be delayed. Because the water used for irrigation in dry areas must be supplied from the water sources, if these water sources do not exist, the approach already discussed here will not be useful. In dry lands, it is necessary to consider a water utilization system that is similar to a closed water recycling system.

In an ordinary natural cycle, precipitation is recycled as shown in Fig. 1 [3]. Part of the water penetrates the soil and is absorbed by plants and is then evaporated from the above-ground plants. As long as this flow is guaranteed, plants grow normally. The amount of evaporated water reaches as high as tens to hundreds of liters for a large tree. In contrast,

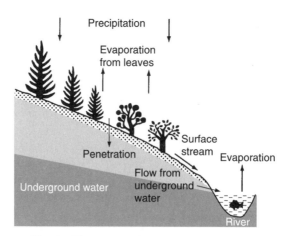

**Fig. 1**  The route of precipitated water.

water fixed through the photosynthesis process is $<1\%$ and more than 95% of the water merely passes through the tree's roots. The amount of evaporated water (uptake) depends on the difference between the maximum moisture content and current moisture content of the air [4].

Even in dry lands, the humidity is at least 20–30% and sometimes reaches as high as over 60%. The temperature is $>50°C$ and the difference between the highest and lowest temperatures in a day is 20–40°C. For example, if the air at 50°C and with 30% humidity is cooled to 10°C at night, $15.5\,\text{g/m}^3$ of water will be condensed. Thus, utilizing such water effectively is an important approach in any area where deforestation is in progress.

## 6   STIMULI-RESPONSIVE POLYMER GELS

The abundant energy in the desert and drylands is solar, which can be used as a source of heat and electrical energy. It may be possible to utilize such unused resources to recycle water. Stimuli-responsive gels are one of the materials that may have the potential to achieve this goal. Study of these materials has increased in recent years.

There are many natural hydrogels such as konnyaku, agar, and kamaboko. As polymer chemistry develops, synthetic polymer gels are increasingly used for separation, and as medical materials and sealants. However, these gels do not respond to external stimuli. Since the discovery and theoretical development of the volumetric phase transition by Tanaka (MIT) [9] many stimuli-responsive gels have been synthesized and their stimuli responsive behavior has been studied.

Osmotic pressure $\pi$, which controls the swelling and shrinking of gels, can be expressed by the following equation [10]:

$$\pi = vkT\left[\frac{\phi}{2\phi_0} - \left(\frac{\phi}{\phi_0}\right)^{1/3}\right] - \frac{\Delta F\phi^2}{v} - \frac{kT[\ln(1-\phi)+\phi]}{v} + fvkT\left(\frac{\phi}{\phi_0}\right)$$

As $\pi = 0$ at equilibrium,

$$\tau = 1 - \frac{\Delta F}{kT} = vv\frac{[2(\phi/\phi_0)^{1/3} - (2f+1)(\phi/\phi_0)]}{\phi^2} + 1 + \frac{2}{\phi} + \frac{2\ln(1-\phi)}{\phi^2}$$

where $v$ is the number of polymer molecules per unit volume, $k$ is the Boltzmann constant, $T$ is the absolute temperature, $\phi$ is the volume fraction of the polymer chain, $\Delta F$ is the free energy based on the interaction between the polymer and solvent, $v$ is the volume per one solvent molecule, $f$ is the number of dissociated counter ions per chain between crosslink points, and $\tau$ is equivalent temperature. From this equation, it can be seen that the degree of swelling of a gel can be controlled by changing the degree of ion dissociation. The influence of the acetone concentration on the volumetric changes of an acrylamide gel has been studied [11]. A discontinuous volumetric phase transition takes place at a constant acetone concentration and its volume can change by as much as 1000 times.

In deserts solar energy is abundant. It can be utilized to provide electrical and thermal energy. Tanaka and other researchers reported on a gel that shows volumetric changes in response to an electrical stimulus. The polymer is made of a partially hydrolyzed acrylamide gel crosslinked by N,N′-methylenebisacrylamide. They reported that when this polymer is placed in a 50% acetone aqueous solution and direct voltage is applied, the gel starts to shrink from the cathode; the entire gel shrunk significantly at $>2.5V$ [9, 10]. Aside from this, there have been many electric field-responsive gels and their stimuli-responsive behavior has been studied [12–15]. If a gel that responds quickly to an electrical stimulus and repeats significant swelling and shrinking can be synthesized, a water recycling system powered by a solar battery is feasible.

A thermoresponsive polymer can also use another form of energy, that is, heat. A thermoresponsive polymer is a polymer that dissolves in water but undergoes phase separation if heated to a certain temperature as a result of dehydration. This polymer possesses a lower critical solution temperature (LCST). The thermoresponsivity of acrylamide-type thermo-responsive polymers is shown in Table 1 [16]. The transition temperature (LCST) ranges from 5–70°C. The transition temperature can be readily controlled by copolymerizing two types of monomers [17] (see Fig. 2).

A thermoresponsive polymer hydrogel can be synthesized by copolymerizing these acrylamide-type monomers and a difunctional monomer such as methylenebisacrylamide. Hirokawa and other researchers studied the temperature dependence of the degree of swelling of a thermoresponsive poly(N-isopropylacrylamide) gel in pure water [18]. It was observed that a discontinuous volumetric phase transition occurred around the LCST of the polymer and the volume change was 8 times.

**Table 1** Phase-transition temperature of acrylamide-type thermoresponsive polymers.

| Polymer | Transition temperature (°C) | Heat of transition (cal/g) |
|---|---|---|
| Poly(N-ethylacrylamide) | 72.0 | – |
| Poly(N-n-propylacrylamide) | 31.0 | 11.6 |
| Poly(N-n-propylmethacrylamide) | 27.0 | 12.9 |
| Poly(N-isopropylacrylamide) | 30.0 | 11.1 |
| Poly(N-isoprolylmethacrylamide) | 43.2 | 12.0 |
| Poly(N-cyclopropylacrylamide) | 45.2 | 3.5 |
| Poly(N-cyclopropylmethacrylamide) | 60.0 | 4.2 |
| Poly(N-ethylmethylacrylamide) | 56.0 | 5.0 |
| Poly(N-N,N'-dimethylacrylamide) | 32.0 | 6.3 |
| Poly(N-acrylpyrroridone) | 56.0 | 1.3 |
| Poly(N-ethylacrylpiperidine) | 5.5 | 10.0 |

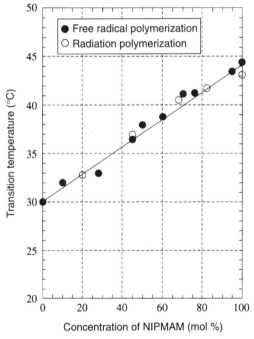

**Fig. 2** Phase transition temperature of the copolymer from a thermoresponsive monomer [17].

## 7 THERMORESPONSIVE POLYMER HYDROGELS

An aqueous solution of poly(vinyl methylether), which is one of the thermoresponsive polymers, dehydrates due to the thermal motion of hydrating water around the methoxy groups and exhibits phase transition (see Fig. 3) [19]. This transition is reversible with respect to the temperature changes, and the transition temperature depends on the concentration of salt in the aqueous solution (see Fig. 4). The poly(vinyl methylether) aqueous solution crosslinks by irradiating x-rays or an electron beam and forms a hydrogel. Similar to the aqueous solution, this gel also shows thermoresponsivity. It swells and shrinks depending on temperature. Its volume changes are illustrated in Fig. 5 [20].

Up to a transition at 38°C, the volume decreases proportionately with temperature, but it shows constant value above the transition temperature. This phenomenon is thermally reversible. The gel, barely heated above the transition temperature, contains water. Figure 6 depicts the effect of salts on the equilibrium swelling of the gel.

The volume does not change with the nitrate ion but with an iodide ion it swells. On the other hand, volume shrinks with chlorine, sulfate, and phosphate ions [21]. The effect of anions on the swelling behavior of gels relates to the ratio $(n/r_1)$, the valence $n$ of the anion, and radius of the ion

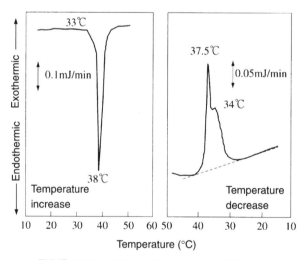

PVME concentration: 5wt%, scan rate: 20°C/min

**Fig. 3** The DSC thermograms of poly(vinyl methylether) aqueous solution [19].

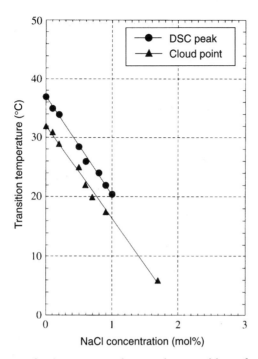

**Fig. 4** Influence of salt concentration on the transition of poly(vinyl methyl-ether) aqueous solution [19].

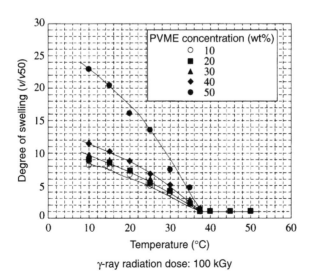

γ-ray radiation dose: 100 kGy

**Fig. 5** Temperature dependence of the swelling of poly(vinyl methylether) gel.

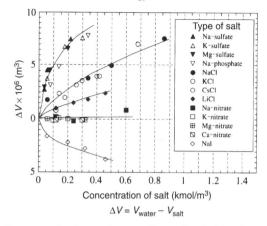

**Fig. 6** Influence of salt on the swelling of poly(vinyl methylether) gel.

$r_1$. When this gel is used to absorb and desorb water, the response time is an important factor.

According to Tanaka and others the response time is proportional to the square of the representative size of the gel. Figure 7 shows two gels with different structures prepared by changing the temperature during synthesis. Their thermoresponsivity is also shown in Fig. 8. A gel with a fine sponge structure showed fast response times of 30 s for swelling and 100 for shrinking [22]. A fibrous gel with a 200 μm diameter showed an extremely fast response time of 0.1 s for both swelling and shrinking (see Fig. 9) [23].

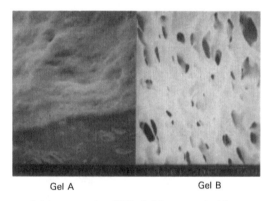

Gel A                    Gel B

Gel A: prepared at 23°C; Gel B: prepared while
the temperature is rising

**Fig. 7** Scanning electron microscope (SEM) photomicrographs of poly(vinyl methylether) gel.

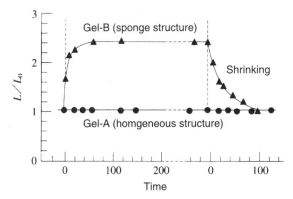

$L_0$ is the length at 40°C; $L$ is the length at time $t$.
The size of the swollen gel was that of a 1 cm cube

**Fig. 8**   Thermoresponsivity of poly(vinyl methylether) gel.

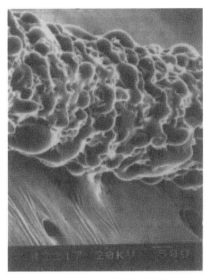

**Fig. 9**   A SEM photomicrograph of fibrous gel [23].

## 8   PROPOSAL FOR NEW WATER RECYCLING SYSTEMS

As just described, 95% of the water absorbed by plants simply passes through the plant and evaporates into the air. Thus, polymer gels, which absorb and desorb water depending on temperature, are proposed to be used to recycle water prior to its evaporation into the air.

Conditions are assumed to be:

1)  Daytime temperature: 45°C, humidity: 40%,
    Water content in air: 26.8 g/m$^3$ air;
2)  Night-time temperature: 10°C, humidity: 100%,
    Water content in air: 9.4 g/m$^3$ air; and
3)  Annual precipitation equivalence: 760 mm/yr, 2 mm/d,
    2000 cm$^3$/m$^2$ (2000 g/m$^2$)

[Calculation]

Amount of condensed water: $26.8 - 9.40 = 17.4$ g/m$^3$ air

Amount of recovered air: 2000 g/17.4 g/m$^3$ $= 115$ m$^3$

Recovery time: 10 h

Air circulation rate: 0.192 m$^3$/min/m$^2$ $= 19$ cm/min.

In this case, the recovered water per 1 m$^2$ is calculated. Because the device requires the same area as an actual cultivation area if this calculation were to be used, a device with 100× the capacity was considered. Thus, the air circulation rate of the module when the contact area is increased to 100 m$^2$ is 30 cm/s. Based on these results and the properties of the gel, it requires a 3 cm thick fibrous gel assuming the use of the fastest responding fibrous gel (with a response time of 0.1 s). If the amount of water absorbed is 2000 g/d, and the packing fraction of the gel in the air is 30%, the space to store the gel per cultivation area of 1 m$^2$ is 6 l. A schematic diagram of this system is depicted in Fig. 10. In the desert, the temperature decreases by radiant cooling at night and the difference between daytime and night-time temperatures can sometimes be as much as 40°C. When the temperature decreases below the dewpoint water that becomes fog has a temperature below the phase-transition temperature of the gel. A hydrophilic gel absorbs water and the gel swells. At sunrise, the temperature increases and the water-containing polymer gel is heated above the transition temperature. Due to phase transition, the water is expelled by the shrinking force and is then supplied to plants. This system can be operated by utilizing the differential energy due to the difference between daytime and night-time temperatures.

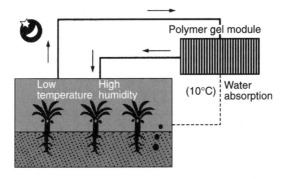

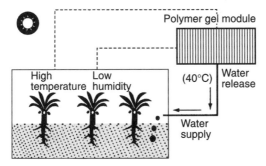

**Fig. 10**   Water recovery and reuse system using a thermoresponsive polymer gel.

The reforestation system proposed here works similarly to the way a greenhouse does. As a variation, a system has been considered in which a thermoresponsive gel is placed near the roots of plants. When the temperature is high and evaporation is rapid during the daytime, the gel gradually and passively releases water. Then, the gel absorbs water from the air during the night-time when temperature falls and stores it for the next day. With this system, ease of operation and sustaining water for the field are also expected.

# 9   CONCLUSIONS

The proposed small trial system and the formidable size of an actual desert provide research with a tremendous challenge. However, in the Namibian desert, a Gomimushidamashi, which is a desert bug, digs a thin trench on the surface of the desert, stands across the trench facing in the direction of the wind, and obtains water by condensing the moisture in the air onto its fine body hairs. It is a natural wonder that gives researchers pause for thought.

## REFERENCES

1   Akagi, H. (1990). *The Nature and Life of Desert*, Chijin Shobo, p. 208.
2   Ishi, H. (1990). *Report on Earth Environments*, Iwanami Shinsho, p. 132.
3   Nakano, H. *et al*. (1989). *Forest and Science*, Tokyo Shoseki, p. 44.
4   Toyama, M. (1989). *Challenge to the Forestation of Desert*, Yomiuri Shinbun-sha.
5   Flory, P.J. (1959). *Polymer Chemistry*, vol. II, K. Kanamaru and S. Oka, Transl., Maruzen, p. 529.
6   Masuda, F. (1990). *Chemical Review: Organic Polymer Gels* (8), p. 52.
7   Toyama, M. (1990). *Zosui Gijutsu* **16**: 35.
8   Toyama, M. (1987). *Fine Chemicals*, pp. 1, 3 and 5.
9   Tanaka, T. *et al*. (1982). *Science* **218**: 467.
10  Tanaka, T. (1986). *J. Phys. Soc., Jpn.* **41**: 542.
11  Hirokawa, Y. *et al*. (1984). *Microbial Adhesion and Aggregation*, New York: Springer-Verlag, p. 177.
12  Shiga, A. *et al*. (1989). *Kobunshi Ronbunshu* **46**: 709.
13  Osada, Y. *et al*. (1987). *Polym. Preprints, Jpn.* **37**: 3116.
14  Maekawa, S. *et al*. (1988). *Polym. Preprints, Jpn.* **38**: 494.
15  Maekawa, S. *et al*. (1988). *Polym. Preprints, Jpn.* **38**: 3901.
16  Itoh, S. (1989). *Kobunshi Ronbunshu* **46**: 473.
17  Itoh, S. (1984). *Inst. Textile Polym. Res. Rep.* **144**: 13.
18  Hirokawa, Y. *et al*. (1984). *J. Chem. Phys.* **81**: 6379.
19  Hirasa, K. (1984). *Inst. Textile Polym. Res. Rep.* **144**: 69.
20  Hirasa, K. (1987). *Proc. 2nd Polym. Gel Symposium*, Soc. Polym. Sci., Jpn., p. 19.
21  Huang, X. *et al*. (1988). *J. Chem. Eng., Jpn.* **21**: 10.
22  Hirasa, K. (1986). *Kobunshi* **35**: 1100.
23  Hirasa, K. *et al*. (1989). *Kobunshi Ronbunshu* **46**: 661.

# CHAPTER 3

# Sanitary Products and Environmental Problems

*SEIRO NISHIO*

## Chapter contents

## 1   INTRODUCTION

If environmental problems are to be understood in the context of evaluating the effect on our living environment and the preservation of nature, sanitary products are unique in a sense that they are misunderstood from both points of view. In particular, the disposable diaper has a short history as a commercial "diaper". There has been unreasonable criticism and doubt from mothers with regard to the newly developed disposable diapers compared to traditional cloth diapers. To address these concerns we will discuss environmental problems associated with sanitary products.

## 2   TERMINOLOGY

### (1)   Definition of Sanitary Products

If hygiene means "paying attention to daily life, keeping our surroundings clean, maintaining our health, and protecting ourselves from illness" (Daijirin, Sansei-do), the fields covering sanitary products should be broad. However, as it is widely understood that sanitary products mean sanitary napkins and disposable diapers, this chapter will adopt this definition.

### (2)   Definition of Environmental Problem by Sanitary Products

### (a)   *Effect on daily environment*
Problems in air pollution include water pollution, the unpleasant odors that occur during the process of rubbish collection, intermediate treatment (burning or crushing) and final treatment (dumping).

### (b)   *Problems on resources and preservation of nature*

1)   Problems related to the preservation of resources (including recycling); and
2)   Problems related to ecology and scenery.

## 3   MATERIALS USED FOR SANITARY PRODUCTS

The materials used for sanitary products are defined in the Sanitary Products Standard (Ministry of Health, No. 285) and the Self-imposed Standard for Sanitary Products in the Law of Pharmacy. Disposable diapers are manufactured according to the same standards, and the contents used are almost the same (see Table 1).

**Table 1** Constituents for sanitary products.

| Components | Surface material | Absorbent |
|---|---|---|
| Water repellent | 1. Organic polymers (plastics) | Synthetic fiber nonwoven cloth |
| | | Open pore plastic sheet |
| | 2. Natural or modified fibers | Cotton or rayon nonwoven cloth |
| | 3. Composites | Mixed nonwoven cloth of natural, modified, and synthetic fibers |
| Fixation material | 1. Natural fibers | Cotton-like pulp (pulp fibers are disentagled like cotton) |
| | 2. Modified fibers | Crepe paper (including tissue paper) |
| | | Rayon-cotton |
| | 3. Organic polymers | Superabsorbent polymers (SAP) |
| | 4. Composites | |
| | 4-1. Natural fiber + organic polymer | Sheet containing SAP |
| | 4-2. Natural fiber + porous material | Deodorant sheet (including activated charcoal) |
| Adhesive | 1. Organic polymers | Polyethylene sheet |
| | | Polymeric elastic sheet or water-repellent nonwoven cloth |
| | 2. Natural fibers | Water-repellent paper |
| | 3. Composites | Polylaminate |
| Elastic material | 1. Organic polymers | Adhesive tape |
| | | Debonding agent |
| | 2. Natural fibers | Tape, string etc. |
| Identification material | Organic polymers | Hot melt |
| | | Cold glue |
| Material | 1. Natural rubbers | Rubber string |
| | 2. Organic polymers | Synthetic rubber string |
| | | Elastic polymer film (including urethane) |
| | | Synthetic fiber elastic nonwoven cloth |
| Material form and name | 1. Natural or modified fibers | Colored cotton thread or rayon thread |
| | | Colored synthetic fiber thread or colored plastic sheet |

The ratio between natural material and organic polymer for sanitary products

| Name of sanitary product | Natural material | Organic polymer | Total |
|---|---|---|---|
| Sanitary napkin | 55%±5% | 45%±5% | 100% |
| Disposable diaper | 60%±5% | 60%±5% | 100% |

**Table 2(a)** Production of disposable diapers.

Unit: t, 1000 pieces

| | Disposable | | | | | | | | | | | |
| | For adults | | | | | | | | | | | |
| | Panty-type | | | | Flat-type | | | | Others (pads) | | | |
| Unit | 1000 pieces | Ratio with the previous year (%) | t | Ratio with the previous year (%) | 1000 pieces | Ratio with the previous year (%) | t | Ratio with the previous year (%) | 1000 pieces | Ratio with the previous year (%) | t | Ratio with the previous year (%) |
|---|---|---|---|---|---|---|---|---|---|---|---|---|
| 1994 Jan–Mar | 45,512 | 112 | 5,714 | 117 | 104,742 | 103 | 7,053 | 104 | 129,477 | 141 | 4,531 | 152 |
| Apri–Jun | 48,187 | 108 | 6,337 | 118 | 108,506 | 102 | 7,361 | 103 | 139,324 | 116 | 4,904 | 127 |
| Jul–Sept | 49,189 | 94 | 6,365 | 102 | 105,464 | 104 | 7,166 | 103 | 155,850 | 140 | 5,497 | 141 |
| Oct–Dec | 60,409 | 117 | 7,998 | 128 | 113,047 | 98 | 7,696 | 99 | 181,052 | 137 | 6,184 | 135 |
| Yearly total | 203,297 | 107 | 26,404 | 116 | 431,759 | 102 | 29,276 | 102 | 605,703 | 133 | 21,116 | 140 |
| 1995 Jan–Mar | 64,456 | 142 | 8,390 | 147 | 103,313 | 99 | 6,993 | 99 | 167,701 | 130 | 5,696 | 126 |
| Apri–Jun | 59,182 | 123 | 7,624 | 120 | 105,672 | 97 | 7,109 | 97 | 170,027 | 122 | 6,208 | 127 |
| Jul–Sept | 57,908 | 118 | 7,264 | 114 | 106,571 | 101 | 7,197 | 100 | 174,528 | 112 | 6,431 | 117 |
| Oct–Dec | 72,887 | 121 | 8,960 | 112 | 110,725 | 98 | 7,389 | 96 | 208,168 | 115 | 7,298 | 118 |
| Yearly total | 254,433 | 125 | 32,238 | 122 | 426,281 | 99 | 28,688 | 98 | 729,424 | 119 | 25,633 | 121 |

# 4 PRODUCTION OF SANITARY PRODUCTS AND ITS CONSEQUENCE AS RUBBISH

## (1) Production

The data on production of sanitary products are shown in Table 2.

## (2) Amount of Rubbish

### (a) Sanitary napkins

The average weight of a sanitary napkin is approximately 6 g, and the weight of blood and sweat per napkin is about 2.5 g. Thus, the weight of the napkins dumped per year is approximately 85,000 metric tons.

### (b) Disposable diapers

Diapers for babies and adults are classified differently. In the future, the production of baby diapers and flat-type adult diapers is not expected to increase. However, a twofold increase in production is predicted for panty-type and pad-type adult diapers (see Table 3).

### (c) Fraction of sanitary products in the domestic rubbish

Because the annual domestic waste is about 50 million tons, the sanitary napkin at 0.17% and the disposable diaper at 2.4%, combine for a total of 2.6% of domestic waste.

**Table 2(a)**   Continued.

Unit: t, 1000 pieces

| Disposable | | | | | | | | | | | |
| For adults | | | | For babies | | | | Total for adults and babies | | | |
| Total for adults | | | | | | | | | | | |
| 1000 pieces | Ratio with the previous year (%) | t | Ratio with the previous year (%) | 1000 pieces | Ratio with the previous year (%) | t | Ratio with the previous year (%) | 1000 pieces | Ratio with the previous year (%) | t | Ratio with the previous year (%) |
|---|---|---|---|---|---|---|---|---|---|---|---|
| 279,731 | 120 | 17,298 | 118 | 1,361,263 | 111 | 61,720 | 108 | 1,640,994 | 113 | 79,018 | 110 |
| 296,017 | 109 | 18,602 | 113 | 1,363,068 | 104 | 62,604 | 102 | 1,659,085 | 105 | 81,206 | 104 |
| 310,503 | 117 | 19,028 | 113 | 1,349,718 | 110 | 61,416 | 110 | 1,660,221 | 111 | 80,444 | 110 |
| 354,508 | 118 | 21,858 | 117 | 1,481,187 | 104 | 65,065 | 100 | 1,835,695 | 107 | 86,933 | 104 |
| 1,240,759 | 116 | 76,796 | 115 | 5,555,236 | 107 | 250,805 | 105 | 6,795,995 | 109 | 327,601 | 107 |
| 335,470 | 120 | 21,079 | 122 | 1,389,022 | 102 | 58,090 | 94 | 1,724,492 | 105 | 79,169 | 100 |
| 334,881 | 113 | 20,941 | 113 | 1,450,512 | 106 | 62,752 | 100 | 1,785,393 | 108 | 83,693 | 103 |
| 339,007 | 109 | 20,892 | 110 | 1,440,608 | 107 | 61,356 | 100 | 1,779,615 | 107 | 82,248 | 102 |
| 391,780 | 111 | 23,647 | 108 | 1,489,976 | 101 | 63,709 | 98 | 1,881,756 | 103 | 87,356 | 100 |
| 1,401 | 138 | 86,559 | 113 | 5,770,118 | 104 | 245,907 | 98 | 7,171,256 | 106 | 332,456 | 101 |

The number of diapers was announced from April 1990.
Three classifications for adults were announced from January 1993.
Reference: Federation of Japan Sanitary Material Industries. 3-36-12, Takada, Toyoshima-ku, Tokyo 171, Japan.

# 5   UNIQUENESS OF SANITARY PRODUCT WASTE FROM THE LEGAL POINT OF VIEW

The content of waste is clearly classified and stated for domestic waste in the general waste category of the "Law on the Treatment of Waste and Cleaning, 1970, No. 137. Modified in 1993. Hereinafter called Law."

The disposable diaper itself fits the definition of waste from a legal point of view, although the diaper consists of multiple materials. However, the used diaper is a combination of materials used for the diaper and human waste. Thus, such a combination does not belong strictly to the definition of domestic waste. Thus, the used diaper is not legally recognized as domestic waste. This is unique only for disposable diapers.

By law, domestic waste from collection to final treatment is the responsibility of individual local communities.

However, as the used diaper is not in the domestic waste category, it can be interpreted that the local community has no legal obligation for the collection and final treatment of used diapers. As a solution to this problem, the consensus of the diaper industry was to convince the general public that the diapers were a daily necessity. However, what they

**Table 2b**  Production and sales of sanitary napkin.

Unit: 1000 yen

| Product name | | | January–December, 1994 | | | | |
|---|---|---|---|---|---|---|---|
| | | | Amount | Ratio against previous year | Number | Ratio against previous year | Average unit price |
| Paper, cotton-like pulp, staple cotton products | Production | | 39,407,986 | 99 | 5,623,033 | 98 | 7,01 |
| | Sales | Domestic | 36,621,989 | 101 | 5,395,425 | 99 | 6,79 |
| | | Export | 2,500,894 | 95 | 179,039 | 91 | 13,87 |
| | End of the month inventory | | 2,425,811 | – | 266,397 | – | – |
| Paper, cotton-like pulp products | Production | | 5,498,288 | 94 | 819,385 | 90 | 6,71 |
| | Sales | Domestic | 5,511,427 | 94 | 821,765 | 91 | 6,71 |
| | | Export | 7,045 | 99 | 1,777 | 99 | 3,96 |
| | End of the month inventory | | 89,61 | – | 12,532 | – | – |
| Paper products | Production | | 4,912,508 | 102 | 1,021,341 | 110 | 4,81 |
| | Sales | Domestic | 3,628,206 | 108 | 776,622 | 116 | 4,67 |
| | | Export | 1,193,158 | 83 | 221,801 | 86 | 5,38 |
| | End of the month inventory | | 161,370 | – | 37,390 | – | – |
| Cotton-like products | Production | | 8,085,923 | 104 | 2,192,426 | 106 | 3,69 |
| | Sales | Domestic | 7,008,547 | 100 | 1,945,053 | 103 | 3,60 |
| | | Export | 1,021,804 | 125 | 231,400 | 131 | 4,42 |
| | End of the month inventory | | 172,922 | – | 47,577 | – | – |
| Other products | Production | | 174,380 | 96 | 18,271 | 108 | 9,54 |
| | Sales | Domestic | 173,112 | 96 | 18,100 | 108 | 9,56 |
| | | Export | – | – | – | – | – |
| | End of the month inventory | | 5,925 | – | 662 | – | – |
| Subtotal | Production | | 58,079,085 | 99 | 9,674,466 | 100 | 6,00 |
| | Sales | Domestic | 52,943,281 | 101 | 8,956,965 | 100 | 5,91 |
| | | Export | 4,722,901 | 96 | 634,017 | 100 | 7,45 |
| | End of the month inventory | | 2,855,644 | – | 364,558 | – | – |

Final product, sanitary products, 1000 pieces
Except for the products with degreased cotton

Reference: Ministry of Health

**Table 3** Amount of disposable diapers as waste.

| Items | For adults | | | | For babies | Total disposable diapers |
|---|---|---|---|---|---|---|
| Style | Panty-type | Flat-type | Pads | Total for adults | Million pieces | – |
| Weight per product | 126.7 g | 67.3 g | 35.7 g | – | 42.6 g | – |
| Average water uptake per product | ≈ 250 g | ≈ 150 g | ≈ 70 g | – | ≈ 120 g | – |
| Average weight as waste per product | ≈ 377 g | ≈ 217 g | ≈ 106 g | – | ≈ 163 g | |
| Annual production | 254.4 Million pieces | 426.3 Million pieces | 720.4 Hundred pieces | – | 5,770 Million pieces | – |
| Annual quantity of waste | ≈ 96,000 t | ≈ 93,000 t | ≈ 76,000 t | ≈ 260,000 t | ≈ 940,000 t | ≈ 1,200,000 t |

Caution 1. Weight per product and annual production is the data for 1995.
Caution 2. Water uptake per product, weight per product as waste. The annual amount of waste is an estimate.

succeeded in doing was teaching consumers how to dispose of the used diapers in order to avoid contamination of the waste collectors.

This situation was finally solved in June 1986 when Tokyo announced that nine items including disposable diapers and sanitary napkins would be collected and treated as burnable waste. Subsequently, a law was enacted October 1, 1986.

Disposable diapers and sanitary napkins were recognized formally as domestic waste for the first time (see Tables 4 and 5).

## 6  INFLUENCE ON DAILY ENVIRONMENT

### (1)  Collection

Due to the aforementioned problems a specialized collection vehicle for the used diapers was tested in order to avoid contamination of the waste collectors. The following are the results of the trials (Table 6).

### (a)  *Situation during loading*

The waste that was loaded was later subjected to more pressure by the rotating blade of the collection vehicle and many of the large trash bags

**Table 4** Analysis of waste.

| Items | Classification | Flammable | Non-flammable | Estimated daily amount | Estimated annual amount | | | Explanation | Reference sources |
|---|---|---|---|---|---|---|---|---|---|
| | | | | | Nationwide | Ratio % | Municipality | | |
| 1. Paper containers | Milk container | ○ | | 26 | 112,110 | 7.13 | 7,993 | The paper used is polyethylene laminated paper. There | Milk: Statistics on daily products |
| | Juice container | | | 6 | 25,260 | 7.13 | 1,801 | is no problem associated with the polymer as the polyethylene film | Japanese Federation of Juice |
| | Sake container | | | 1 | 4,450 | 7.53 | 335 | is very thin. Although there are some containers with aluminum foil, no | Sake: Food Statistics Monthly |
| | Soy milk container | | | 1 | 4,060 | 7.13 | 289 | problems are encountered as the aluminum foil is also thin. | |
| | Total | | | 34 | 145,880 | | 10,418 | | |
| 2. Disposable diapers | | ○ | | 20 | 105,110 | 5.85 | 6,149 | There are also diapers made of synthetic fibers (nonwoven cloth), cotton-like pulp, and polyethylene film. The used diapers contain human waste and thus need to be handled in a sanitary manner. It is virtually impossible to remove polypropylene or polyester nonwoven cloth and polyethylene and to separate flammable materials from nonflammable materials. | Japan Association of Sanitary Material Industries Social Welfare Investigation, Tokyo Investigation of Aged Population Living in Kita-ku |
| 3. Sanitary products | | ○ | | 16 | 65,630 | 7.42 | 4,870 | Polyester nonwoven cloth or modified cotton is used. It is necessary to be handled in a sanitary manner. In practice, they are regarded as flammable waste. There is no problem associated with their quantities. | Japanese Association of Sanitary Material Industries |
| 4. Cards, catalogues, record jackets | | ○ | | Unknown | | | | Sometimes the used coating paper and dye are unknown. However, the amount is negligible. However, cards made of plastics, record jackets and the plastic wrappings are separable waste. | |

| Item | | | | | | | Description | Source |
|---|---|---|---|---|---|---|---|---|
| 5. Commercial paper garbage bags | ○ | | 4 | 12,561 | 8.79 | 1,04 | There are waxed papers and polyethylene laminated papers. However, no problems associated with these materials are known. | Annual Report on Home Economics |
| 6. Mixed woven cloths | ○ | | Unknown | | | | Mixed woven materials with polyester and nylon. There is no problem with excessive heat generation | |
| 7. Chemical cleaning cloths | ○ | | Unknown | | | | They are made of 15% nylon and 85% rayon. Similar to the mixed fibers, there is no problem. | |
| 8. Tobacco filters | ○ | | 4 | | | 1,123 | The raw material of the filter is poly(vinyl acetate) and there is no problem. | |
| 9. Cotton made of modified fibers | ○ | | 4 | 16,381 | 7.13 | 1,168 | Same as the mixed fiber cloths. | Annual Statistics of Fibers |
| 10. Plastic bags | | ○ | 156 | 662,700 | 7.13 | 47,251 | They are plastic bags. Even when "nontoxic" designation is used, there is a problem of excessive heat generation After the waste is treated, the waste in the container will be separated. | National Guild of Polyethylene Film Industries |
| 11. Portable waste bags | | ○ | | | | | | |
| 12. Portable heaters | | ○ | 5 | 21,000 | 7.13 | 1,497 | The content content is mainly iron powder, water, activated charcoal, and salt. The packaging material is made of modified fibers. | Japanese Association of Disposable, Portable Heater Industries |

(Caution 1) Amount per capita is according to the national survey conducted in 1980. Department of City Beautification, Tokyo (June 23, 1986).

**Table 5** Chemical components of different physical appearance of domestic waste (dry material basis).

(Department of Beautification, Tokyo. 1983).

| Physical makeup | Classification | Ash % | Flammable components % | Main component of flammable materials % | | | | | | | |
| --- | --- | --- | --- | --- | --- | --- | --- | --- | --- | --- | --- |
| | | | | Carbon % | Hydrogen % | Volatile chlorine % | Flammable sulfur % | Exotherm | | Oxygen % | Nitrogen % |
| | | | | | | | | High value kcal/kg | Low value kcal/kg | | |
| Flammable | Paper | 8.22 | 91.77 | 43.84 | 6.18 | 0.301 | 0.052 | 4.227 | 3.893 | 41.12 | 0.27 |
| | Fibers | 2.40 | 97.59 | 50.83 | 6.33 | 0.252 | 0.084 | 4.879 | 4.537 | 36.99 | 3.10 |
| | **** | 16.76 | 83.23 | 41.22 | 5.64 | 0.163 | 0.096 | 4.063 | 3.759 | 33.54 | 2.56 |
| | Plants | 10.23 | 89.76 | 45.57 | 5.79 | 0.217 | 0.049 | 4.438 | 4.152 | 37.29 | 0.83 |
| Inappropriate for burning | Plastics | 5.09 | 94.90 | 70.51 | 8.34 | 3.398 | 0.048 | 13,538 | 7.900 | 11.70 | 0.39 |

**Table 6** The test results by a waste collection vehicle.

| | |
|---|---|
| 1. Objectives | To evaluate various problems during waste collection by adhering simulated human waste on a disposable diaper for a baby, simulating several ways to dispose of the waste, loading and unloading by an actual waste collection vehicle. |
| 2. Participants | Eight companies that form the Technology Committee of the Japan Federation for Sanitary Material Industries as well as one company that has obtained a certificate from Tokyo as a waste collection contractor. |
| 3. Test location | Old Paper Storage, Edogawa Plant, Honshu-seishi Ltd. |
| 4. Vehicle used for the test | Backmaster 2t, volume 4.2 m$^3$ |
| 5. Test date | September 2, 1986 (although another test also was conducted in May, it has the appearance of a pre-test. Thus, the results are not shown here) |
| 6. Experimental methods | (1) Materials<br>  (a) Disposable diaper, M size for babies<br>  (b) Additives |

| | | | |
|---|---|---|---|
| | (i) Simulated stool | Materials | Weight ratio |
| | (For soft stool) | CMC sodium | 1 g |
| | | Talc (particle diameter less than 30 μm) | 60 g |
| | | Methylene blue | 12 mg |
| | | Ion-exchanged water | 60 g |
| | (ii) Simulated urine | Materials | Composition |
| | | Saline solution | 10 *l* |
| | | Red dye No. 2 | 1 g |

(2) Samples (disposable diapers with simulated stools or urine)

| | | | | Symbols |
|---|---|---|---|---|
| (a) Simulated stool | | Weight | Treatment | A |
| | | 100 g (i) | Folding only | Ⓐ |
| | | (ii) | Folded and taped | B |
| (b) Simulated urine | | 200 g (i) | Folding only | Ⓑ |
| (Assuming nighttime) | | (ii) | Folded and taped | C |
| (c) Simulated urine | | 100 g (i) | Folding only | Ⓒ |
| (Assuming daytime) | | (ii) | Folded and taped | |

(*Continued*)

**Table 6**  Continued.

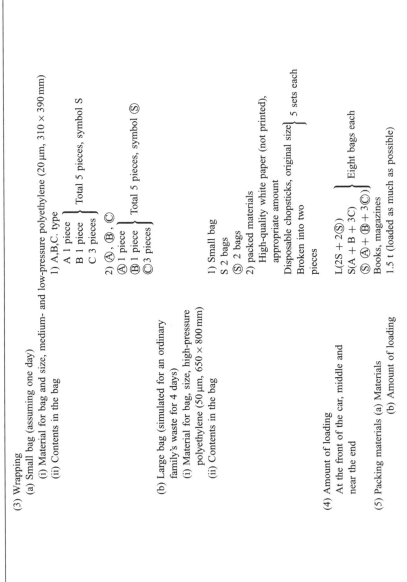

(3) Wrapping
(a) Small bag (assuming one day)
  (i) Material for bag and size, medium- and low-pressure polyethylene (20 μm, 310 × 390 mm)
  (ii) Contents in the bag

    1) A,B,C. type
      A 1 piece
      B 1 piece ⎱ Total 5 pieces, symbol S
      C 3 pieces ⎰

    2) Ⓐ, Ⓑ, Ⓒ
      Ⓐ 1 piece
      Ⓑ 1 piece ⎱ Total 5 pieces, symbol Ⓢ
      Ⓒ 3 pieces ⎰

(b) Large bag (simulated for an ordinary family's waste for 4 days)
  (i) Material for bag, size, high-pressure polyethylene (50 μm, 650 × 800 mm)
  (ii) Contents in the bag

    1) Small bag
      S 2 bags
      Ⓢ 2 bags
    2) packed materials
      High-quality white paper (not printed), appropriate amount
      Disposable chopsticks, original size ⎫
                                   ⎬ 5 sets each
      Broken into two pieces ⎭

(4) Amount of loading
At the front of the car, middle and near the end

  L(2S + 2Ⓢ)
  S(A + B + 3C) ⎱ Eight bags each
  Ⓢ(Ⓐ + Ⓑ + 3Ⓒ) ⎰
  Books, magazines

(5) Packing materials (a) Materials
                 (b) Amount of loading

1.5 t (loaded as much as possible)

(6) Loading method

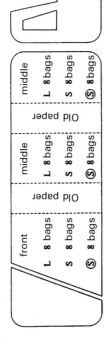

| front | Old paper | middle | Old paper | middle |
|---|---|---|---|---|
| L 8 bags | | L 8 bags | | L 8 bags |
| S 8 bags | | S 8 bags | | S 8 bags |
| Ⓢ 8 bags | | Ⓢ 8 bags | | Ⓢ 8 bags |

(7) Staying time in the waste collection vehicle
After loading, the rubbish were taken out after a certain time
(8) Method of checking sample conditions
Eyeball inspection

failed (hereinafter abbreviated as L). The following are the causes for the failure: 1) The air content in "L" was high due to the packing of high-quality papers; and 2) pointed objects such as disposable chopsticks were present in "L."

### *(b)  Situation during unloading*

1) Despite the fact that the bags were loaded in the front, middle section, and rear section, the separation of these three sections was not clear at the time of unloading.

2) The relationships between the position of loading and failure:

    i)    There were no failed bags in the "L" that were loaded in the deeper section. However, the majority of bags in the middle section and near the front failed.

    ii)    Regardless of the loading position, there was no failure of small bags (hereinafter abbreviated as "$\text{Ⓢ}$").

    iii)    Although there were no failed "$\text{Ⓢ}$," which were loaded near the front, there were many bags that showed leakage of simulated human waste inside the bag. However, there was no leakage outside of the bag. This is considered to be due to the influence of the pressure caused by the rotating blade.

3) Difference by the Treatment of Disposable Diapers

    Under any conditions, whether it is "L" or "$\text{Ⓢ}$," as far as the deformation and leakage of simulated human waste are concerned, Ⓐ, Ⓑ and Ⓒ were superior to A, B and C. Therefore, if consumers follow the Ⓐ, Ⓑ and Ⓒ methods, there is no need to be concerned with contamination of waste collectors (readers are referred to the (2) Sampling in the Experimental Section of the Rubbish Collection Car Project, 6, for the designation of Ⓐ, Ⓑ and Ⓒ, and A, B and C).

### (2)  Intermediate Treatment

### *(a)  Suitability for burning*

The recognition of sanitary napkins and disposable diapers as flammable materials is the result of testing using a large-scale furnace. Even if the 3300 municipalities in Japan decided to use cooperative facilities, it is impossible to install large-scale furnaces similar to those in large cities. On the other hand, over 80% of the babies in Japan are now using disposable diapers. Furthermore, in this aging society, the increase in the use of diapers by adults has doubled. Thus, if it is feasible to burn used

**Table 7** Chemical analysis of physically different domestic wastes (wet basis).

| | Water % | Flammable % | Ash % | Carbon % | Hydrogen % | Oxygen % | Nitrogen % | Sulfur | | | Chlorine | | | Exotherm | |
| --- | --- | --- | --- | --- | --- | --- | --- | --- | --- | --- | --- | --- | --- | --- | --- |
| | | | | | | | | Flammable % | Non-flammable % | Total % | Volatile % | Residue % | Total % | High value kcal/kg | Low value kcal/kg |
| Paper | 23.5 | 70.64 | 6.30 | 33.74 | 4.76 | 31.66 | 0.20 | 0.040 | 0.040 | 0.079 | 0.231 | 0.418 | 0.649 | 3252 | 2995 |
| | 4.15 | 4.38 | 1.20 | 1.96 | 0.24 | 2.30 | 0.07 | 0.032 | 0.013 | 0.033 | 0.118 | 0.377 | 0.335 | 175 | 162 |
| Fibers | 10.30 | 87.61 | 2.08 | 45.71 | 5.68 | 33.08 | 2.83 | 0.076 | 0.028 | 0.105 | .220 | 0.595 | 0.814 | 4392 | 4085 |
| | 4.75 | 5.95 | 1.48 | 5.79 | 0.51 | 2.83 | 2.22 | 0.054 | 0.012 | 0.053 | 0.142 | 1.242 | 1.196 | 700 | 687 |
| ??? | 78.04 | 18.14 | 3.81 | 8.97 | 1.22 | 7.33 | 0.56 | 0.020 | 0.059 | 0.079 | 0.034 | 0.910 | 0.944 | 887 | 821 |
| | 4.12 | 3.19 | 1.95 | 1.58 | 0.21 | 1.39 | 0.11 | 0.010 | 0.066 | 0.070 | 0.012 | 0.712 | 0.719 | 162 | 152 |
| Plants | 41.37 | 52.42 | 6.19 | 26.63 | 3.37 | 21.80 | 0.47 | 0.030 | 0.052 | 0.081 | 0.122 | 1.461 | 1.583 | 2601 | 2419 |
| | 20.66 | 17.54 | 4.62 | 9.03 | 1.10 | 7.29 | 0.14 | 0.017 | 0.011 | 0.024 | 0.033 | 1.183 | 1.189 | 926 | 867 |
| Plastics | 20.73 | 75.24 | 4.02 | 55.80 | 6.65 | 9.27 | 0.31 | 0.038 | 0.012 | 0.050 | 3.159 | 7.566 | 10.725 | 6232 | 5872 |
| | 8.65 | 8.45 | 0.73 | 5.46 | 2.91 | 2.87 | 0.13 | 0.012 | 0.002 | 0.013 | 1.280 | 3.176 | 3.379 | 432 | 423 |

N = 6, upper row: average value, lower row: standard deviation
Institute for Cleaning, Tokyo (1978)

**Table 8** Chemical analysis of physically different domestic wastes (dry basis).

| | Flammable % | Ash % | Carbon % | Hydrogen % | Oxygen % | Nitrogen % | Sulfur | | | Chlorine | | | Exotherm | |
| --- | --- | --- | --- | --- | --- | --- | --- | --- | --- | --- | --- | --- | --- | --- |
| | | | | | | | Flammable % | Non-flammable % | Total % | Volatile % | Residue % | Total % | High value kcal/kg | Low value kcal/kg |
| Paper | 91.77 | 8.22 | 43.84 | 6.18 | 41.12 | 0.27 | 0.052 | 0.051 | 0.103 | 0.301 | 0.538 | 0.839 | 4227 | 3893 |
| | 1.64 | 1.64 | 0.63 | 0.14 | 1.37 | 0.09 | 0.042 | 0.018 | 0.044 | 0.154 | 0.476 | 0.412 | 76 | 69 |
| Fibers | 97.59 | 2.40 | 50.83 | 6.33 | 36.99 | 3.10 | 0.084 | 0.032 | 10.116 | 0.252 | 0.639 | 0.891 | 4879 | 4537 |
| | 1.88 | 1.88 | 4.63 | 0.39 | 6.17 | 2.35 | 0.056 | 0.015 | 0.055 | 0.181 | 1.333 | 1.279 | 593 | 593 |
| ??? | 83.23 | 16.76 | 41.22 | 5.64 | 33.54 | 2.56 | 0.096 | 0.292 | 0.388 | 0.163 | 4.052 | 4.215 | 4063 | 3759 |
| | 7.38 | 7.38 | 4.40 | 0.73 | 3.05 | 0.33 | 0.046 | 0.358 | 0.383 | 0.080 | 2.702 | 2.742 | 349 | 317 |
| Plants | 89.76 | 10.23 | 45.57 | 5.79 | 37.29 | 0.83 | 0.049 | 0.098 | 0.148 | 0.217 | 3.119 | 3.336 | 4438 | 4125 |
| | 5.99 | 5.99 | 3.26 | 0.37 | 2.62 | 0.16 | 0.016 | 0.42 | 0.041 | 0.050 | 2.970 | 3.016 | 123 | 110 |
| Plastics | 94.90 | 5.09 | 70.51 | 8.34 | 11.70 | 0.39 | 0.048 | 0.015 | 0.063 | 3.898 | 9.641 | 13.538 | 7900 | 7450 |
| | 0.95 | 0.95 | 2.35 | 3.58 | 3.57 | 0.12 | 0.016 | 0.003 | 0.018 | 1.325 | 4.034 | 4.063 | 421 | 510 |

N = 6, upper row: average value, lower row: standard deviation
Institute for Cleaning, Tokyo (1978)

sanitary napkins and disposable diapers by a small furnace that can be installed even by small municipalities and institutions, the benefits will be incalculable. Unfortunately, there are technical difficulties in using small furnaces to burn used disposable diapers.

In recent years, the improvement in the quality of disposable diapers has been remarkable. In particular, the increased water uptake is remarkable and the increased water uptake is equivalent to increased water content. The tables shown here are comparative studies on dry and wet waste conducted by the Institute for Cleaning, Tokyo (see Tables 7 and 8).

Readers are referred to the column of high water content. The wet materials show extremely low exotherms in comparison to the dry material. Needless to say, the low exotherm indicates low burning temperature.

The water content of the used disposable diaper is estimated to be around 50 to 60%. However, the water content at the center of the diaper is much higher than these values and is greater than the example mentioned in the foregoing. The higher the water uptake, the lower the flammability. Experts predicted that the high water-content disposable diapers will show incomplete burning near the center of the diaper, thus the incompletely burned materials must be removed each time in the case of the batch-type furnace. The results obtained by the Japan Federation of Sanitary Material Industries on the burning of wet diapers are shown in what follows. The test was conducted in cooperation with the Japan Association of Production Machine Industries, Inc., and the measurement was done by the Association of Mechanical and Electrical Testing (all the burning tests were done in the same manner (see Fig. 1, Tables 9 to 13)). It was found that, even using a commercially available small furnace, the used disposable diapers were successfully burned. It is anticipated that there will be no problem in burning sanitary napkins.

### (b) Evaluation of air pollution

In order to correct the misconception that disposable diapers utilize a large amount of plastics and thus generate toxic gases upon burning, and also to evaluate the amount of air pollution when disposable diapers and sanitary napkins are burned by a small furnace, the following tests were conducted during burning and the results are shown in Tables 14 and 15.

Although the furnace used for the test had a small bed area and was not subjected to laws regarding air pollution, it nonetheless passed the air pollution standards. The reason why Example 1 is omitted from the test

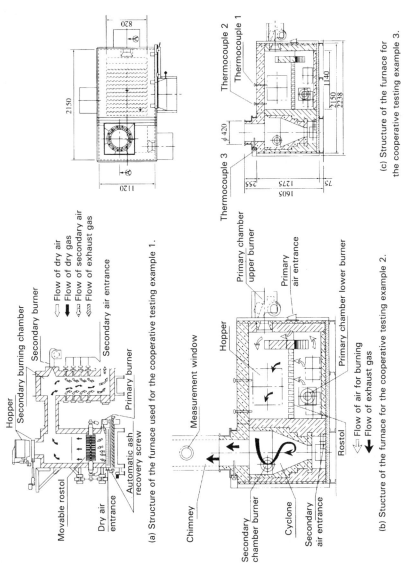

**Fig. 1** Structure of the furnace for cooperative testing.

(a) Structure of the furnace used for the cooperative testing example 1.

(b) Stucture of the furnace for the cooperative testing example 2.

(c) Structure of the furnace for the cooperative testing example 3.

Hopper
Secondary burning chamber
Secondary burner
Secondary air entrance
Primary burner
Automatic ash recovery screw
Dry air entrance
Movable rostol

⇨ Flow of dry air
■ Flow of dry gas
⇨ Flow of secondary air
⇨ Flow of exhaust gas

Measurement window
Chimney
Secondary chamber burner
Cyclone
Secondary air entrance

Thermocouple 2
Thermocouple 1
Thermocouple 3
Primary chamber upper burner
Primary air entrance
Hopper
Primary chamber lower burner
Rostol

⇨ Flow of air for burning
■ Flow of exhaust gas

**Table 9**   Summary of the furnace for the cooperative testing

| | The cooperative testing furnace No. 1 | The cooperative testing furnace No. 2 | The cooperative testing furnace No. 3 |
|---|---|---|---|
| Treatment capacity | $40 \times 10^4$ kcal/h | 55–65 kg/h | 60–65 kg/h |
| Furnace type | Burning by gasification (day) | Forced heating | Forced heating |
| Burning area (m²) | 1.0 | 0.74 | 0.76 |
| Burning chamber volume (m³) | 2.8 | 1.13 | 1.81 |
| Air circulation system | Forced air | Natural air circulation | Forced air |
| Secondary burning facility | Two burners | Three burners | One burner |
| Exhaust gas treatment | Gasification, cyclone | Repeated burning, cyclone | Gravity type |
| Test date | October 4, 1988 | October 18, 1988 | October 25, 1988 |

**Table 10**   Measurement items and methods.

| Measured items | Measurement location | Measurement frequency | Measurement method | Standard |
|---|---|---|---|---|
| Properties of waste | Once/day | 70°C × 1 week (moisture) | Environmental Protection No. 95 | |
| Exotherm | | Once/day | Pump-type calorimeter | Environmental Protection No. 95 |
| Elemental analysis | | Once/day | Elemental analyzer | Environmental Protection No. 95 |
| Temperature of the furnace | Primary chamber temperature | Thermocouple | K-TYPE thermocouple (JIS C 1602) | JIS Z 8808 |
| Exit temperature of primary chamber | Temperature of the furnace | Continuous | K-TYPE thermocouple (JIS C 1602) | JIS Z 8808 |
| (Secondary chamber temperature) | Thermocouple | Continuous | K-TYPE thermocouple (JIS C 1602) | JIS Z 8808 |
| Weight of waste loaded | | As necessary | Manual measurement (a balance) | Minimum measure less than 0.1 kg |

**Table 11**   Components of total solids of the cooperative burning test material.

| Component items | Unit | Example 1 | Example 2 | Example 3 |
|---|---|---|---|---|
| Carbon | % (DB) | 45.20 | 52.03 | 49.96 |
| Hydrogen | % (DB) | 7.10 | 8.07 | 7.65 |
| Nitrogen | % (DB) | 0.10 | 8.07 | 7.65 |
| Oxygen | % (DB) | 40.26 | 34.21 | 34.99 |
| Sulfur | % (DB) | < 0.01 | < 0.01 | < 0.01 |
| Chlorine | % (DB) | 0.12 | 0.17 | 0.12 |
| Ash | % (DB) | 7.21 | 4.62 | 6.95 |

Caution 1. The material employed for the cooperative burning test was used disposable diapers for adults collected from hospitals and institutions.

**Table 12**   The physicochemical properties of burning test materials.

| Items | Unit | Example 1 | Example 2 | Example 3 |
|---|---|---|---|---|
| Type of disposable diapers | | Panty type, flat type | Panty type, flat type | Flat type |
| Water content | % | 48.8 | 50.9 | 58.8 |
| Total solid | % | 51.2 | 43.1 | 41.2 |
| Ash | % (DB) | 7.21 | 4.62 | 6.95 |
| Flammable component | % (DB) | 92.79 | 95.38 | 93.05 |
| High exotherm | kcal/kg (DB) | 4.650 | 5,310 | 5,010 |
| Low exotherm | kcal/kg (DB) | 4.270 | 4,870 | 4,600 |

Caution 1. The material employed for the cooperative burning test was used disposable diapers for adults collected from hospitals and institutions.

results is because the wood used as co-fuel had a preservative and the measured value exhibited abnormal results.

## (c)   *Bad odor*

From the measurements for the four components of bad smells caused by human waste, all the components exhibited below-standard results (see Table 16).

## (d)   *Dioxin*

(i)   Summary of dioxin

*What is dioxin?* Dioxin is the collective name of 75 compounds called polychlorodibenzo-p-dioxin and 135 compounds called polychlorodibenzofuran. They are toxic but the toxicity of 2,3,7,8-TCDD is the highest.

**Table 13** Results on the suitability for burning.

| | Example 1 | Example 2 | Example 3 |
|---|---|---|---|
| States at the time of loading the furnace | Mixed woods per paper box | Per plastic bag | Per plastic bag |
| Coverage of human waste (%) | P type 44.4<br>F type 56.0 | P type 52.2<br>F type 59.4 | F type 59.2 |
| Hh (kcal/kg·WB) | 2.380 | 2.290 | 2.060 |
| Hl (kcal/kg·WB) | (Mixed burning 3080) | 1,760 | 1,540 |
| Amount of treatment per hour (kg/h) | 101 = 44.1 + 56.9<br>(56% supplemental wood fuel) | 59.4 | 62.1 |
| Supplemental fuel type (density) | Kerosene (0.8) | Kerosene (0.8) | Heavy oil (0.86) |
| Consumption per hour (l.h) | 38.3 | 18.9 | 15.4 |
| Exhaust gas temperature, primary (°C) | 941 | 803 | 757 |
| Average temperature, secondary (°C) | 1,007 | 870 | 700 |
| Chimney temperature (°C) | 100 | — | 383 |
| Residue of burning | | | |
| Weight of ash (kg) | 7 | 1.8 | 4.9 |
| Weight ratio to the rubbish (%) | 2.77 | 0.9 | 2.0 |
| Weight reduction (%, DB) | — | 2.1 | 2.8 |
| Appropriateness for burning | Can be burned with mixed woods | Nearly completely burned | Nearly completely burned |

Caution
P type: used disposable diaper for adults (panty type)
F type: used disposable diaper for adults (flat type)
Hh: High exotherm per wet basis
Hl: Low exotherm per wet basis (the value in the parenthesis in Example 1 is based on the mixed fuel assuming the heat generation of wood to be 4000 kcal/kg)
WB: wet base

**Table 14** Measured items and measurement methods.

| | Exhaust gas | Place of measurement | Measurement frequency | Measurement methods | Standard used |
|---|---|---|---|---|---|
| **Measured items** | Exhaust gas temperature | Measurement duct | Once/continuous | K-TYPE thermocouple (JIS C 1602) | JIS Z 8808 |
| | Amount of exhaust gas | Measurement duct | Once/day | Pitot tube (JIS B 8330) | JIS Z 8808 |
| | Dust | Measurement duct | Once/day | Tubular filter paper method (moving sampling method) | JIS K 0107 |
| | Hydrogen chloride | Measurement duct | Once/day | Mercury thiocyane (II) absorption spectroscopy | JIS K 0103 |
| | Sulfur oxides | Measurement duct | Once/day | Relative turbidity method | JIS K 0104 |
| | Nitroxides | Measurement duct | Once/day | JIS B 7982 chemical luminescence method (moving method) | |
| | Oxygen | Measurement duct | Once/day | JIS B 7983 magnetic wind method of Sylconi method | |
| | Carbon dioxide | Measurement duct | Continuous | Olzat gas chromatography | |
| | Carbon monoxide | Measurement duct | Continuous | Olzat gas chromatography | |
| | Ammonia | Measurement duct | Once/continuous | Indophenol absorption spectroscopy | JIS K 0099 |
| | Acetaldehyde | Measurement duct | Once/continuous | Gas chromatography | Environmental Protection Agency Announcement No. 9 |
| | Trimethylamine | Measurement duct | Once/day | Gas chromatography | Environmental Protection Agency Announcement No. 9 |
| | Sulfur componds | Measurement duct | Once/day | Gas chromatography | Environmental Protection Agency Announcement No. 9 |

**Table 15** Results from the analysis of exhaust gas components.

| Measured items | Example 2 | Example 3 | General restriction values |
|---|---|---|---|
| Amount of wet gas ($Nm^3/h$) | 680 | 1.080 | |
| Amount of dry gas ($Nm^3/h$) | 610 | 980 | |
| Dust concentration ($g/Mm^3$) | 0.042 (0.03) | 0.10 (0.15) | < 0.5 |
| Oxygen concentration (%) | 8.6 | 14.7 | |
| Nitrogen dioxide (12% $O_2$) (ppm) | 86 | 76 | < 250 |
| Hydrogen chloride (12% $O_2$) (ppm) | 6.2 | 26 | < 430 |
| Sulfur oxides (ppm) | < 2.4 | 40 | $K$ value restriction |
| Ammonia (ppm) | 3.8 (2.8) | 4.2 (6.8) | |
| | 29.6 (51.2) | | |
| Hydrogen sulfide (ppm) | 0.019 (0.017) | 0.092 (0.14) | |
| Trimethylamine (ppm) | < 0.0005 | < 0.0005 | |

Caution 1. The value in the parenthesis is the average value of 12% $O_2$ equivalence during the measurement time zone.
Caution 2. The ammonia in Example 2 was measured twice.

**Table 16** Comparison to the regulatory standards of $SO_x$, the components of bad odor.

| Regulatory standards | Equivalence items | Example 2 | Example 3 | Regulation values |
|---|---|---|---|---|
| Sulfur oxides | $K$ value equivalence | < 0.01 | 0.50 | ≤ 1.17 |
| Ammonia | Cm equivalence (ppm) | 0.00034 | 0.00049 | 1–5 |
| | | | 0.034 | |
| Hydrogen sulfide | Cm equivalence (ppm) | $1.7 \times 10^{-6}$ | $1.1 \times 10^{-5}$ | 0.02–0.2 |
| Trimethylamine | Cm equivalence (ppm) | $< 4.4 \times 10^{-3}$ | $< 5.8 \times 10^{-2}$ | 0.005–0.07 |

Caution. $K$ equivalence value of sulfur oxides was calculated by the following equation regulated by law:
$K = q \times 10^3/He^2$
Legend of symbols
  $K$: local values determined by law
  $q$: the amount of sulfur oxides ($Nm^3/h$)
  $He$: the corrected height of the exhaust exit (m)

*The mechanism of dioxin production* Although it is not completely understood, the following cases might be important.

1) Production mechanism during chlorine bleaching:

- Lignin that is contained in pulp reacts with chlorine to be used for bleaching.

- Chlorinated phenol that is contained in the original wood condenses during decomposition.
- Impurities such as dibenzofuran and dibenzoxin are chlorinated during bleaching. (T. Wakimoto, LAPHA, May, 1995 with permission).

2) Production mechanism during burning of waste:

- Reactions between aromatic compounds and chlorine.
- Gas–solid phase reaction under the coexistence of hydrocarbon and chlorine.
- Dimerization in the 300–350°C temperature range.
- The dioxin, which is contained in the waste, enters into the exhaust gas.

3) Other production mechanisms of dioxin:

- Forest fires; volcano eruptions
- Exhaust gas from automobiles; natural production by insecticides and weedkillers.

### (ii)  History of the discovery of dioxin production during chlorine bleaching

The United States (or US) Environmental Protection Agency (EPA) discovered a minute quantity of dioxin for the first time in the world in a factory wastewater at a paper and pulp plant.

### (iii)  Concentration of dioxins

In 1984, the Ministry of Health, Japan, announced that the influence of dioxin on humans is 100 pg/kg weight/day using 2,3,7,8-TCDD as a representative example. Six years later, the Japan Federation of Paper Industries reported the results of the analysis of dioxin concentration of pulp and paper. Judging from these values, it became clear that the concentration of dioxin in pulp and paper products is extremely small (see Table 17).

### (iv)  Toxicity of dioxin

Dioxin attracted attention as a carcinogen. The dioxin problem with regard to paper and pulp first became serious in the United States. However, the US Federal Drug Administration (FDA) later testified at a Congressional Subcommittee on Health and Environment that as far as the available information is concerned the extremely small amount of dioxin cannot

**Table 17** An example of dioxin concentration analysis

|  | Pulp | High-quality paper |
|---|---|---|
| Maximum value | 1.8 ppt | 0.71 ppt |
| Minimum value | 0 | 0.01 |
| Median | 0.54 | 0.58 |

Japan Federation of Paper Industries (1990).

pose a serious threat to human beings. Additionally, the US Consumer Products Safety Committee (CPSC) reported that dioxin in paper products can be ignored (announcement date unknown). Subsequently, there has been no discussion of these harmless paper product theories in the United States or any other countries.

(v)   Reaction from the paper and pulp industries

Efforts to change from chlorine bleaching to oxygen bleaching have resulted in nearly a 100% changeover to the oxygen bleaching system in the main paper-producing countries.

(vi)   Countermeasures for dioxin formation during waste incineration

When tests were first conducted on small-scale furnaces, the interest in dioxin by the general public was minimal. Although there was an emergency regarding many suspected pollutants, dioxin was not included in the tests. Fortunately, the formation mechanism of dioxin has almost been clarified by experts on incineration and its countermeasures are currently being sought. It is desired that the nontoxicity of dioxin will be proven quantitatively in the near future as was the case for paper. For reference, Table 18 lists the "Regulation on the dioxin concentration during incineration of general waste" in major Western countries and Japan.

## (3)   Final Treatment

The preservation of water quality in land reclamation is an important subject. For this, three approaches were taken.

### (a)   Measurement on related items during incineration test

Considering that the final use of waste is in land reclamation, dissolution of incinerated ash and its content was measured and the results are shown in Table 19. This study reveals no problem with used disposable diapers.

**Table 18** Regulation of the dioxin concentration in the exhaust gas of general waste incineration.

| Country, Area | | Regulation values and conditions |
|---|---|---|
| Japan | 0.5 ng-TEQ/Nm³ | Expected value of newly installed furnace (1990), Decree of the Head of Environmental Department, Ministry of Health (Environmental Decree No. 260). |
| EEC proposal (1992) | 0.1 | |
| Germany | 0.1 | |
| The Netherlands | 0.1 | |
| Austria | 0.1 | |
| Sweden | 0.1 | Newly installed furnace |
| | 0.5–2 | Already installed furnace |
| Denmark | 1.0 | |
| United Kingdom | 1.0 | (The target value is 0.1 ng – TEQ/Nm³) |
| Norway | 2 | |
| Italy | (0.01 mg/Nm³) | As PCDDs + PCDFs |
| | (0.05 µg/Nm³) | As TCDDs + TCDFs |
| United States | 5–30 ng/Nm³ | As PCDDs + PCDFs, newly installed furnace with treatment capacity of more than 250 t/day |
| | 75 ng/Nm³ | As PCDDs + PCDFs, newly installed furnace with treatment capacity of less than 250 t/day |
| | 3–50 ng/Nm³ | As PCDDs + PCDFs, already installed furnace with treatment capacity of more than 2200 t/day |
| | 125 ng/Nm³ | As PCDDs + PCDFs, already installed furnace with treatment capacity of 2200 to 250 t/day |
| | 500 ng/Nm³ | As PCDDs + PCDF, already installed furnace with less than 250 t/day. |
| Belgium | – | |
| Portugal | – | |
| France | – | |
| Ireland | – | |
| Spain | – | |

Countries with regulations determined by law are Germany, The Netherlands, United Kingdom and Austria. The other countries only have guidelines.

*1. From the reference of EEC in December, 1992

*2. In Japan, the Ministry of Health announced in January 1997 that the expected value of a newly installed furnace will be less than 0.1 ng. The Environmental Protection Agency is in the process of determining guidelines for dioxin concentration in exhaust gas based on the air pollution protection law.

**Table 19**  Measurement items and methods.

| | Measurement items | Measurement location | Measurement frequency | Measurement methods | Applicable regulations |
|---|---|---|---|---|---|
| Dissolution test (Incinerated ash) | Total mercury | | Once/day | Atomic absorption spectroscopy | Decree by Environmental Protection Agency No. 13 |
| | Cadmium | | Once/day | Atomic absorption spectroscopy | Decree by Environmental Protection Agency No. 13 |
| | Lead | | Once/day | Atomic absorption spectroscopy | Decree by Environmental Protection Agency No. 13 |
| | Arsenic | | Once/day | Absorption spectroscopy using silver diethyldithiocarbamic acid | Decree by Environmental Protection Agency No. 13 |
| | Alkyl mercury | | Once/day | Gas chromatography | Decree by Environmental Protection Agency No. 13 |
| | Organic phosphate | | Once/day | Gas chromatography | Decree by Environmental Protection Agency No. 13 |
| | PCB | | Once/day | Gas chromatography | Decree by Enviornmmetal Protection Agency No. 13 |
| | Hexavalent chromium | | Once/day | Absorption spectroscopy using diphenylcarbazide | Decree by Environmental Protection Agency No. 13 |
| | Cyan | | Once/day | Absorption spectroscopy using 4-pyridinecarboxylic acid-pyrazon | Decree by Environmental Protection Agency No. 13 |
| | Copper | | Once/day | Atomic absorption spectroscopy | Decree by Environmental Protection Agency No. 13 |
| | Zinc | | Once/day | Atomic absorption spectroscopy | Decree by Environmental Protection Agency No. 13 |
| | Flourine | | Once/day | Absorption spectroscopy using lantan.alizalincomblexison | Decree by Environmental Protection Agency No. 13 |
| | Organic chlorine | | Once/day | Absorption spectroscopy | Decree by Environmental Protection Agency No. 13 |
| Contents | Water content | | Once/day | Gravimetry | Decree by Environmental Protection Agency No. 13 |
| | Reduction upon heating | | Once/day | Gravimetry | Sedimentation measurement method |
| | Total memory | | Once/day | Atomic absorption spectroscopy | Sedimentation measurement method |
| | Cadmium | | Once/day | Atomic absorption spectrscopy | Sedimentation measurement method |
| | Lead | | Once/day | Atomic absorption spectroscopy | Sedimentation measurement method |
| | Arsenic | | Once/day | Absorption spectroscopy using diethylthiocarbamin | Sedimentation measurement method |

## (b) Microanalysis of heavy metals that are contained in the constituents of disposable diapers

Although there may be a minute amount, nonetheless there are heavy metals and toxic components in nature.

Tables 21 and 22 introduce the literature from "heavy metals in" and "toxic materials in local soils" reported by The Institute for Cleaning, Tokyo, in 1978 and 1980. From these references, analysis of contents of the materials used for disposable diapers was deemed necessary. Thus, microanalysis of heavy metals that are contained in the constituent materials of disposable diapers was requested by the Japan Environmental Health Center, Inc., and the results are shown in Table 23 (February 1986).

## (c) Literature search

Using a database, a search of the literature from Western countries was conducted where the use of disposable diapers was pioneered and land

**Table 20**  Analysis of the incinerated ash components (examples 2 and 3).

|  | Items |  | Concentration | Land reclamation standards |
|---|---|---|---|---|
| **Dissolution test** | Total mercury | mg/l | ≪0.005 | < 0.005 |
|  | Cadmium | mg/l | ≪0.01 | < 0.3 |
|  | Lead | mg/l | ≪0.1 | < 3 |
|  | Arsenic | mg/l | 0.9–0.11 | < 1.5 |
|  | Alkyl mercury | mg/l | ≪0.0005 | Nondetectable |
|  | Organic phosphate | mg/l | ≪0.1 | < 1 |
|  | PCB | mg/l | ≪0.0005 | < 0.003 |
|  | Hexavalent chromium | mg/l | ≪0.04 | < 1.5 |
|  | Cyan | mg/l | ≪0.1 | < 1 |
|  | Copper | mg/l | ≪0.1 | (<3) |
|  | Zinc | mg/l | 0.02–0.4 | (<5) |
|  | Fluorine | mg/l | 0.1–0.53 | (<15) |
|  | Organic chlorine | mg/kg | ≪3 | <40 |
| **Contents** | Water content | w/w% | < 0.5 | – |
|  | Reduction upon heating | w/w% (D.B) | 2.1–2.8 | <15 |
|  | Total mercury | mg/kg (D.B) | < 0.01 | – |
|  | Cadmium | mg/kg (D.B) | 0.22–0.79 | – |
|  | Lead | mg/kg (D.B) | 1.45–5.8 | – |
|  | Arsenic | mg/kg (D.B) | 1.41–2.45 | – |

Caution 1. The concentration shown as 1 ≪ j indicates the lower detection limit (JIS K 0102)
Caution 2. The values in the parenthesis in the land-reclamation standards are reference values (Water Polution Protection Law)

**Table 21** Influence of the heavy metals that are contained in hand-separated domestic rubbish.

| Samples | Details | | CD % | Pb | Zn | Cu | Cr | Hg | As |
|---|---|---|---|---|---|---|---|---|---|
| | Main body | Adhered materials | (μg/g wet) | (μg/g wet) | (μg/g wet) | (μg/g wet) | (μg/g wet) | (μg/g wet) | (μg/g wet) |
| Grains | Rice ~80, bread ~20 | Sand, paper, others | 76.1 (0.159) | 64.0 (0.577) | 62.2 (7.89) | 85.3 (2.48) | 58.7 (0.313) | 55.2 (0.00165) | 75.8 (0.0412) |
| Vegetables | Daikon leaves ~40, onion ~10, pickles of daikon ~10, daikon ~10, potato ~10, leaves of bamboo shoot ~1, others | Sand, paper, others | 63.5 (1.38) | 58.5 (5.07) | 56.3 (31.09) | 27.6 (1.49) | 48.5 (1.68) | 58.2 (0.00436) | 41.2 (0.0241) |
| Fruits | Orange peel ~70, banana skin ~10, grapefruit peel ~5, lemon peel ~5, cherry ~5 | Sand, paper, others | 61.4 (0.093) | 67.8 (0.970) | 42.0 (4.95) | 18.8 (1.26) | 32.0 (0.555) | 17.5 (0.00355) | 40.1 (0.0284) |
| Meat | poke ~60, chicken with bones ~20, salami ~10, ham ~5 | Sand, paper, others | 93.4 (1.38) | 85.4 (5.07) | 83.4 (31.9) | 64.8 (1.49) | 66.5 (1.68) | 9.1 (0.000223) | 56.6 (0.0278) |

Caution 1. The kitchen-related rubbish (340–1,020 g) separated from other domestic rubbish are washed with approximately 500 ml of water carefully. The washed rubbish is called the main body and the water used and the others, such as sand and paper, are called adhered materials.
Caution 2. Upper numbers in each raw is the contribution of the main body to total heavy metals (main body and adgered materials).
Caution 3. The numbers shown are percent value based on the wet.
Caution 4. According to the Origin of Harmful materials in City Rubbish, 1, No. 1 (from the Report of the Institute for Cleaning, Toky (1978)).

**Table 22** Toxic materials that are contained in each district.

| Districts | Moisture dried by air (%) | Reduction by weight by heating (%) | pH | Cd (µg/g) | Pb (µg/g) | Zn (µg/g) | Cu (µg/g) | Cr (µg/g) | Hg (µg/g) | As (µg/g) | T-Cl (µg/g) | S-Cl (µg/g) | $Cl^-$ Percent dissolution (%) | T-$SO_4$ (µg/g) | S-$SO_4$ (µg/g) | $SO_4$ Percent dissolution (%) | Root of 7th power | Root of 7th power of the sum of each average | Distance (km) |
|---|---|---|---|---|---|---|---|---|---|---|---|---|---|---|---|---|---|---|---|
| Northern district (1–8) | 0.83 | 6.02 | 7.24 | 0.37 | 74.8 | 15.9 | 96.1 | 84.7 | 0.140 | 11.1 | 320 | 109 | 62 | 1480 | 800 | 42 | 10.8 | 128 | 4.81 |
|  | 0.46 | 3.63 | 0.14 | 0.35 | 49.6 | 60.5 | 71.7 | 69.5 | 0.117 | 5.63 | 150 | 100 | 26 | 1070 | 1050 | 25 | 4.68 |  | 0.31 |
| Eastern district (9–12) | 0.77 | 8.43 | 7.23 | 0.14 | 37.8 | 86.1 | 32.4 | 41.6 | 0.0514 | 4.53 | 1370 | 1260 | 60 | 2860 | 880 | 32 | 5.46 | 5.49 | 2.81 |
|  | 0.53 | 11.21 | 0.23 | 0.24 | 20.0 | 34.2 | 28.5 | 20.4 | 0.0238 | 2.13 | 2260 | 2310 | 31 | 920 | 750 | 21 | 1.64 |  | 0.58 |
| Western district (13–16) | 1.77 | 8.04 | 7.33 | 0.55 | 62.6 | 112 | 35.5 | 46.6 | 0.208 | 5.39 | 1690 | 1610 | 77 | 810 | 170 | 17 | 8.90 | 9.52 | 4.11 |
|  | 1.65 | 4.23 | 0.17 | 0.15 | 78.2 | 58.7 | 20.7 | 16.5 | 0.177 | 2.66 | 2680 | 2720 | 17 | 910 | 200 | 18 | 4.02 |  | 0.34 |
| Central district (17–22) | 1.29 | 6.73 | 7.47 | 0.31 | 57.8 | 93.3 | 27.5 | 32.0 | 0.158 | 5.19 | 290 | 140 | 49 | 1220 | 350 | 44 | 6.11 | 7.39 | 3.28 |
|  | 1.45 | 4.34 | 0.19 | 0.27 | 73.3 | 47.0 | 18.2 | 20.5 | 0.186 | 2.21 | 100 | 94 | 26 | 750 | 90 | 31 | 2.98 |  | 0.83 |
| Eastern seaboard district (23–28) | 1.12 | 8.96 | 7.83 | 0.09 | 31.6 | 163 | 35.6 | 1.02 | 0.0851 | 5.62 | 920 | 750 | 40 | 3660 | 720 | 17 | 5.94 | 6.91 | 1.24 |
|  | 0.99 | 2.63 | 0.24 |  | 30.2 | 180 | 327 | 1.11 | 0.0329 | 3.57 | 1500 | 1550 | 34 | 2880 | 1220 | 16 | 3.47 |  | 0.50 |
| Western seaboard district (29–33) | 2.07 | 8.31 | 7.30 | 0.09 | 50.0 | 74.7 | 40.2 | 35.2 | 0.0931 | 5.39 | 100 | 50 | 47 | 1160 | 70 | 5.8 | 6.58 | 8.15 | 2.53 |
|  | 1.85 | 4.78 | 0.76 |  | 32.1 | 37.5 | 40.4 | 56.0 | 0.115 | 3.15 | 24 | 5 | 11 | 1460 | 80 | 62 | 4.92 |  | 0.69 |
| No. 14 district (34–39) | 3.37 | 6.04 | 6.25 | 0.09 | 23.0 | 63.6 | 33.6 | 41.4 | 0.0535 | 6.93 | 230 | 60 | 34 | 1220 | 340 | 24 | 4.51 | 4.90 | 0.82 |
|  | 3.30 | 1.70 | 1.39 |  | 9.40 | 18.8 | 37.2 | 18.9 | 0.0321 | 3.28 | 160 | 10 | 24 | 1590 | 460 | 23 | 1.59 |  | 0.53 |
| No. 15 district (40–43) | 2.37 | 5.29 | 5.78 | 0.09 | 30.7 | 89.1 | 26.3 | 35.8 | 0.0622 | 6.76 | 130 | 70 | 60 | 960 | 200 | 24 | 4.76 | 5.03 | 2.70 |
|  | 0.82 | 2.81 | 1.46 |  | 24.4 | 51.3 | 10.7 | 30.3 | 0.0192 | 0.69 | 80 | 20 | 21 | 560 | 170 | 17 | 1.68 |  | 0.29 |
| Near roads with heavy traffic (3–8, 10, 13, 14, 20, 29, 30, 33) | 1.36 | 7.24 | 7.22 | 0.36 | 79.0 | 13.9 | 76.1 | 17.2 | 0.167 | 6.82 | 270 | 140 | 54 | 1300 | 350 | 2710.1 | 12.9 | 4.04 |  |
|  | 1.45 | 4.16 | 0.39 | 0.34 | 50.1 | 62.5 | 65.0 | 35.7 | 0.135 | 3.05 | 140 | 90 | 19 | 1010 | 330 | 17 | 4.61 |  |  |
| Total Koto-ku district (1–43) | 1.66 | 7.00 | 7.09 | 0.19 | 47.6 | 1.08 | 44.9 | 92.8 | 0.106 | 6.71 | 570 | 440 | 54 | 1690 | 470 | 27 | 6.76 | 8.38 | 1.05 |
|  | 1.54 | 4.45 | 0.91 | 0.28 | 46.0 | 82.1 | 45.3 | 20.6 | 0.113 | 3.92 | 1200 | 1220 | 26 | 1670 | 720 | 24 | 3.83 |  | 2.83 |
| Outside of Koto-ku district (44–51) | 4.92 | 13.71 | 6.99 | 0.28 | 69.5 | 16.3 | 59.4 | 48.1 | 0.0962 | 5.89 | 220 | 110 | 67 | 1140 | 190 | 18 | 7.41 | 8.81 | 1.46 |
|  | 4.14 | 8.78 | 0.98 | 0.37 | 49.8 | 98.5 | 34.5 | 46.6 | 0.0753 | 1.65 | 190 | 110 | 32 | 760 | 230 | 21 | 3.41 |  |  |

Caution 1. Upper row: average values, lower row: standard deviation
Inside of the parentheses in the district is the sample number
The Institute of Cleaning, Tokyo (1980)

Table 23  Microanalysis of heavy metals in disposable diapers.

| Sample No. | Samples analyzed | Footnotes | Mercury (mg/kg) | Cadmium (mg/kg) | Lead (mg/kg) | Arsenic (mg/kg) |
|---|---|---|---|---|---|---|
| 1 | Nonwoven cloth | Cloth made of 100% PP or close to it | 0.001 Less than | 0.012 | 1.8 | 0.05 Less than |
| 2 | | Cloth made of more than 50% PET with the addition of ES or NBF | 0.047 | 0.021 | 0.5 | 0.33 |
| 3 | | Cloth made mainly of PET and rayon with some ES and NBF | 0.001 Less than | 0.005 Less than | 0.4 | 1.0 |
| 4 | | Cloth made mainly of rayon and PP with some ES and NBF | 0.001 Less than | 0.011 | 1.0 | 0.21 |
| 5 | Crushed pulp | Uncrushed pulp | 0.64 | 0.005 Less than | 0.2 Less than | 0.05 Less than |
| 6 | Absorbing paper | Crepe paper using NBKP | 0.002 | 0.005 Less than | 0.5 | 0.08 |
| 7 | Polymeric absorbent | Crosslinked poly(acrylic acid) type | 0.001 Less than | 0.038 | 0.2 Less than | 0.05 Less than |
| 8 | | | 0.001 | 0.005 Less than | 0.2 Less than | 0.05 Less than |
| 9 | PE film | High-pressure PE | 0.001 Less than | 0.005 Less than | 0.7 | 0.05 Less than |
| 10 | | High-pressure PE | 0.001 Less than | 0.005 Less than | 0.7 | 0.05 Less than |
| 11 | | Calcium carbonate-filled high-pressure PE | 0.002 | 0.11 | 0.7 | 0.07 |
| 12 | Polymeric absorbent | Made of a plastic | 0.001 Less than | 0.005 Less than | 0.2 | 0.27 |
| 13 | Release tape | Made of a plastic | 0.001 Less than | 0.005 Less than | 0.3 | 0.05 Less than |
| 14 | Target tape | Made of a plastic | 0.001 Less than | 0.005 Less than | 0.2 Less than | 1.2 |
| 15 | Spandex | Polyurethane-type | 0.001 Less than | 0.005 Less than | 0.2 Less than | 0.17 |
| 16 | Hot melt | Polyamide, polyester, polyolefin-type | 0.001 Less than | 0.005 Less than | 0.2 Less than | 0.05 Less than |
| 17 | PE film | Calcium carbonate high-pressure PE | 0.001 Less than | 0.38 | 0.3 | 0.21 |
| 18 | Nonwoven cloth | Cloth made mainly of PP | 0.001 Less than | 0.005 Less than | 0.3 | 0.05 Less than |
| 19 | MIX absorbent | Crushed pulp + polymer absorbent | 0.021 | 0.005 Less than | 0.2 Less than | 0.05 Less than |

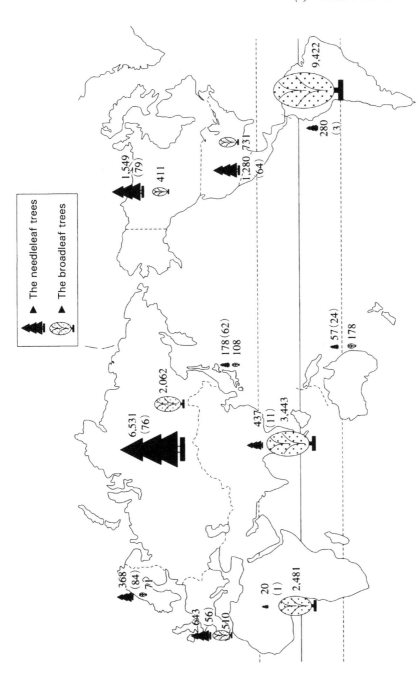

**Fig. 2**   The forest resources in the world (1980).

reclamation is the main mode of waste treatment. The items of interest were:

1) whether or not the work environment is contaminated during waste collection and land reclamation; and
2) whether or not secondary pollution after land reclamation has ever been recorded.

There were no reports on harm to humans, environmental pollution or secondary pollution.

## 7   PRESERVATION OF NATURAL RESOURCES

### (1)   Current State of the Forestry of the World and Japan

#### (a)   Emergency in forestry resources is a problem of the subtropical and southern hemisphere

The emergency situation of forestry resources has not been discussed in a long time. (Readers are referred to Fig. 2 and Table 24.) In the 15 years from 1971 to 1986, 2.9% of the world's forest has disappeared. This trend continues to the present day. Notably, the disappearance of forests in the developing countries is as high as 5.4%. If a simplistic calculation is made, those developing countries will turn into deserts and grasslands in about 280 years.

#### (b)   Current situation in developed countries

It is not possible to discuss the situation in all the developed countries. Those countries that report increases (although minor) in forestry

**Table 24**   The change of forest area in the world in the past 15 years. Units: 10,000 km$^2$.

|  | 1971 | 1986 | Change in 1971–1986 |
|---|---|---|---|
| Developed countries | 1855 | 1859 | +4 |
| Developing countries (tropical countries) | 2345 | 2219 | −126 |
|  | (2052) | (1923) | (−129) |
| World total | 4200 | 4078 | −122 |

According to the "Annual Timber Production" of the Food and Agriculture Organization of the United Nations, the forest areas of developed countries have remained nearly constant (slightly increasing) since 1971.

Caution 1. This forest area is the total of closed forest and rare forest (when observed from the sky, there are few trees. The projected area of the tree to the ground area is at least 10%).

Caution 2. The tropical countries are the areas beween the tropic of Cancer and the tropic of Capricorn.

resources are the United States, Canada and Sweden. These are the main exporting countries of forestry resources and the improvement is due to their forestry improvement policies. These countries, in addition to Japan, which is one of the highest paper-producing countries in the world, have very high paper recycling rates. After Japan (slightly less than 55%), which ranks highest in the world in used paper recycling, the aforementioned countries all show high percentages. Even in pulp materials the concept of recycling is widespread. Sawdust and wood waste are used as chip materials for making pulp. Sometimes, these chips constitute more than 50% of the raw materials for pulp. It is important to recognize that the export of forestry resources of the developed countries is not due to the excess supply, but to the efforts in resource preservation and recycling.

### (c)  Current situation in Japan

When observed from the sky, a forest that makes it difficult to recognize bare ground is called a "closed forest." The ratio of the closed forest to the total area of a country is called a closed forest ratio. The closed forest ratio of Japan is 36%, which far surpasses that of the United States (26%), which is second, and Canada (25%), which is third. However, it should be noted that this high percentage is due to the warm climate and high precipitation as well as the necessity in maintaining the forests to prevent natural disasters. The area of land that can be cultivated in Japan is about 37%. Thus, most of the forests are located in the mountains.

The deregulation of woods and fast economic growth had a major impact on the forests of Japan. The change in lifestyle brought about the reduction of charcoal consumption, and technological advances and change in the industrial sector also reduced the need for wood as railroad supports. The domestic production of wood could not compete with imported wood due to deregulation. From the economic point of view, the drastic reduction of the needs for broadleaf trees could not generate income for labor. The change in construction methods and the materials used as well as imported wood also made the domestic needleleaf trees too expensive.

As a result, the attraction of the forest as an asset is reduced and there are no incentives for investment. As reported by the Forestry Agency, Japanese forestry is at a deficit of 3 trillion yen.

## (2)   The Majority of Pulp Used for Sanitary Products is Imported

Both the sanitary napkin and disposable diaper depend mainly on imported pulp for one simple reason—it is nothing but stable supply and cost. There is too much difference in the production of trees between forestry resources exporting countries and Japan.

The pulp used for sanitary products is from the Japanese trees that are densely planted on mountain slopes, sometimes at an incline of 30 degrees. Due to the conservation of water resources, and once a forest is declared a natural preserve, the freedom to cut trees is severely restricted. These are all reasons for the high cost. By contrast, wood-exporting countries are able to plant trees on flat lands in a less dense manner. Other factors that influence tree cutting are market demands. Thus, as is clear, conditions do not favor Japanese tree production, which is far below the cost performance of other countries.

## (3)   Reason why Used Papers are not Employed for Sanitary Products

### (a)   *The role of pulp and appropriateness of quality*

As long as pulp is used as absorbent, the amount of absorption is an important parameter for evaluation. However, there are other properties that are required for the pulp used for sanitary products.

### (i)   Water absorption and water uptake

When pulp is used for sanitary products, pulp fibers are disentangled and made cottonlike, hence it is called cotton-like pulp (or disentangled pulp).

The water uptake of pulp fiber itself is only about 15 to 20% of its own weight. However, cotton-like pulp holds (not absorbs) a maximum of 17 to 18 times that of its own weight, solely because of the space created by the fibers holding the water. Therefore, the fiber layers (hereinafter called cotton-like pulp layers), which have high porosity, are a better pulp for sanitary products. The requirements for producing high porosity are the following:

1) The length of fiber must be long; and
2) Appropriate hardness (elastic).

### (ii)   Functions other than absorption and holding of water

*Maintenance of shape*   Sanitary products are subject to movement and pressures from various angles, leading to fragmentation of the cotton-like pulp layers. Thus, in order to prevent this the two previously mentioned requirements are necessary.

*Fitting*   Sanitary products are unsuitable if one experiences discomfort when using the product. For the cotton-like pulp, the fluffy appearance is desirable. In addition to absorption and holding of water as well as maintenance of shape, softness is also important, but long fiber length is required.

*Diffusion*   It is desired that the local moisture from the absorption point be quickly dispersed. For this reason, long fiber length is desirable.

(iii)   Types of pulp and evaluation of the fitness of product quality

According to the discussions in (i) and (ii), the fitness of product quality is evaluated on three levels (see Table 25).

From the preceding evaluation, the most important property of sanitary products is water holding, and NBKP, which is known for its high maintenance of shape, is adopted by the majority of manufacturers despite some variation among the origin of the raw materials and producers.

(iv)   Reasons why recycled pulp has poor reputation

1)   It is difficult to require homogeneity in quality for the used paper.
2)   If paper with high quality and good homogeneity is required, it is difficult to guarantee a steady supply of the required amount and thus it becomes expensive.
3)   The used paper undergoes change in quality and degrades after recycling processes. This trend becomes more severe with increased numbers of treatments. Thus, the fiber becomes shorter and more brittle.
4)   The greater the number of recycling processes, the darker in color the fiber becomes.

**Table 25**   Fitness of the product quality based on the different kinds of pulp.

| Required properties | NBKP | NBSP | LBKP | LBSP | TMP | Recycled pulp |
|---|---|---|---|---|---|---|
| Water holding | ○ | △ | △ | × | ○ | × |
| Maintenance of shape | ○ | △ | △ | × | ○ | × |
| Fitting | △ | ○ | △ | △ | △ | × |
| Spreading | ○ | ○ | △ | △ | ○ | × |
| Whiteness | △ | ○ | △ | ○ | × | × |

○ Good; △ medium; × bad.
N, Needleleaf tree; L, broadleaf tree; B, bleaching; K, S, type of chemical treatment; T, chemical treatment; P, pulp

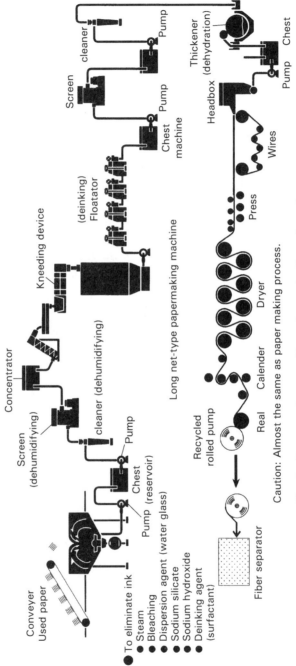

**Fig. 3** Process of cotton-like pulp production of used paper.

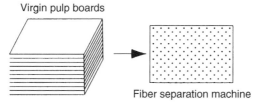

Virgin pulp boards

Fiber separation machine

**Fig. 4**   Manufacturing process of cotton-like pulp from virgin pulp.

## (b)   Recycled pulp is expensive

Readers are referred to Fig. 3 for an explanation of why recycled pulp requires many processes prior to the separation of pulp fibers as is done for cotton-like pulp. Without going through the papermaking process, the separation of fibers cannot be achieved. This is why the recycled fiber is expensive. On the other hand, the most inexpensive absorbent using pulp is cotton-like pulp (see Fig. 4).

## (c)   Problems regarding drug-related law

"Standard for Sanitary Products" (Decree by Minister, No. 285) prohibits used paper to be employed for sanitary products in order to maintain safety and sanitation. A similar self-imposed standard for disposable diapers for babies also prohibits the use of recycled paper. However, because there is no regulation prohibiting the employment of used paper for adult disposable diapers, products employing used paper have been sold until recently; the number has gradually been decreasing in recent years.

## (4)   Approaches for Conservation of Resources

"Wasting of forest resources" is a complaint of many consumers and consumer organizations. This was also the subject of a TV program.

The current situation regarding pulp resources, which is the main raw material for sanitary products, has already been described. However, attention must be given to Fig. 3, where how much energy is consumed from the time the used paper is placed in the machine to the production of recycled rolled panels is shown. Energy consumed is the electricity for operating the machine and the steam to soften and dry the pulp. To generate electricity and to produce steam, petroleum is used. In the case of virgin and recycled pulp, the problem is one of renewable resources. On the other hand, petroleum is a limited-resource fuel.

There is another point worth mentioning. As is clear from Fig. 4, cotton-like pulp does not use any thermal energy. This is the major difference between virgin paper and recycled paper or virgin cotton-like pulp and recycled cotton-like pulp.

Accordingly, there are cases where recycling is not always necessarily "good," although it should not simply be ruled out as the merit of preservation of each resource is different.

## 8   ECOLOGY AND PRESERVATION OF SCENERY

Although it is obvious that sanitary products are consuming forest resources, it has been shown here that their environmental impact has been negligible.

## 9   CONCLUSIONS

This chapter did not address or attempt to solve all the problems regarding sanitary products and the environment. As the aging society continues to grow the consumption of adult disposable diapers will necessarily increase. Thus, there is no guarantee that there will not be additional environmental concerns. There are a number of problems to be studied in the future. This chapter will conclude by briefly describing two of those problems.

### (1)   Weight Reduction of Disposable Diapers

The problems of waste will be more problematic in the 21st century. Of major concern is the total amount of waste. Disposable diapers have successfully reduced the weight per single product and the number of diapers used per day. As a result, a weight reduction of approximately 30% has been achieved. Unfortunately, considering the projected increase in consumption, the aforementioned improvement is still insufficient. However, as long as the weight reduction can be achieved, there are several ways to do so. It can be done either by the improvement in current specifications, development of new materials, or invention of new functions. It is necessary to recognize that these areas need immediate attention.

### (2)   Used Disposable Diapers for Use as Compost

Here, the development of biodegradable polymers is assumed. The importance here is to produce useful materials for plants upon biodegradation. Along with kitchen waste, if compost can be made of disposable diapers, it can contribute significantly to the weight reduction of waste.

# DATA SUMMARY

## Gel compound data index

| | |
|---|---|
| Name | Agarose |
| Classification• network | Natural, organic, polysaccharides |
| Classification• crosslink point | Physical gel |
| Classification• fluid | Hydrogel |
| Manufacturing method | Agar is acetylated or methylated and separated using chloroform into a soluble composition and nonsoluble composition. Agarose can be made by taking out the nonsoluble composition and demethylating or gelling using alkali. It can also be separated by coprecipitating with poly(ethylene glycol). |
| Solvent | Water |
| Characteristics | Agarose gel shows syneresis and hysteresis. Gelation point is approximately $40°C$, the melting point is $90°C$, and molecular weight is around $10^5$. |
| Gel preparation method | Gel is formed when the temperature of heated agarose solution is maintained under $40°C$. |
| Uses | Food. Base material for electrophoresis. Filler for gel filtration. |
| Raw material manufacturer | |
| Product manufacturer | Sigma |
| Related literature | 1) C. Dea, A. A. Mckinnon, and D. A. Rees: *J. Mol. Biol.* **68**: 153 (1972) |
| Chemical formula | |

| | |
|---|---|
| Name | Agaropectin |
| Classification• network | Natural, organic, polysaccharides |
| Classification• crosslink point | Physical gel |
| Classification• fluid | Hydrogel |
| Manufacturing method | Agar is acetylated or methylated and is fractionated using chloroform into a soluble composition and nonsoluble composition. It can be obtained by taking the nonsoluble composition and demethylating or saponifying by alkali. |
| Solvent | Water |
| Characteristics | One of the polysaccharides that makes up agar. Agarose forms a hard gel, but agaropectin has less gel strength as it contains 3 to 10% ester sulfate, around 1% of pyruvic acid and a small amount of hyluronic acid. |
| Gel preparation method | The same method as agarose. |
| Uses | Food additives |
| Raw material manufacturer | |
| Product manufacturer | |
| Related literature | C. Araki: *Japan Chemical Journal* **58**: 1213, 1214, 1362 (1937). |

| | |
|---|---|
| Name | Amylose |
| Classification• network | Natural, organic, polysaccharides |
| Classification• crosslink point | Physical gel |

| | |
|---|---|
| Classification• fluid | Hydrogel |
| Manufacturing method | Create a suspension with 10% agar (w/w) and 13% magnesium sulfate (w/w). Add a small amount of magnesium sulfite and adjust the pH to 6.5 to 7.0, pressurize and heat to 160°C. Cool when the agar has completely dispersed, and maintain 80°C for 25 min, then amylose is precipitated. Gather the sediments while maintaining 80°C and by centrifugal separation (2000 g). After the temperature of the mother liquor drops to 20°C, and then left for several hours at that temperature, amylopectin is deposited. |
| Solvent | Water (pressurized heat), DMSO |
| Characteristics | Molecular weight for linear chain polymer is $5 \times 10^5$–$2 \times 10^6$. Amylose molecules take a helical structure of approximate diameter 1.3 nm that wraps around once with 6 to 7 glucose residues, and an approximate pitch of 0.8 nm. Degree of polymerization 1000–4000 $[\alpha]_D = +220°$ (aqueous solution) |
| Gel preparation method | Heat the solution and cool |
| Uses | Adhesives for cardboard, paint industry, absorbent for fats and oils, food thickening stabilizer gel agent, molecular weight marker |
| Raw material manufacturer | Hayashihara Biochemical Research Center, Nagato Acids |
| Product manufacturer | Biochemical Co. |
| Special note | Store at 4°C |
| Related literature | 1) O. Kjolberg and D. J. Manners: *Biochem. J.* **84**: 50 (1962). 2) T. Kuge and S. Ono: *Bull. Chem. Soc. Japan* **54**: 1264 (1961). Muetgeert: *Advances in Carbohydrate Chemistry*, vol. 16, M. L. Wolfrom, ed., New York: Academic Press (1961), p. 299. |

Chemical formula

| | |
|---|---|
| Name | Amylopectin |
| Classification• network | Natural, organic, polysaccharides |
| Classification• crosslink point | Physical gel |
| Classification• fluid | Hydrogel |
| Manufacturing method | See manufacturing method for amylose. |
| Solvent | Water |
| Characteristics | Molecular weight of linear chain polymer is $15$–$400 \times 10^6$. Relates to agar paste formation. Average chain length 18–24. $[\alpha]_D = +150°$ (in sodium hydroxide solution). |
| Gel preparation method | Heat solution and cool. Gel strength is weaker compared to amylose. |
| Uses | Food, pastes |
| Raw material manufacturer | |

Academic Press
*A Harcourt Science and Technology Company*
525 B Street, Suite 1900, San Diego, California 92101-4495, USA
http://www.academicpress.com

Academic Press
Harcourt Place, 32 Jamestown Road, London NW1 7BY, UK
http://www.academicpress.com

Library of Congress Catalog Card Number: 00-107106

International Standard Book Number: 0-12-394690-5 (Set)

PRINTED IN THE UNITED STATES OF AMERICA
00  01  02  03  04  05  IP  9 8 7 6 5 4 3 2 1

| | |
|---|---|
| Product manufacturer | Nakarai Tesque |
| Related literature | 1) C. T. Greenwood: *Starke* **12**: 169 (1960). |
| | 2) K. Freudenberg and Y. H. Boppel: *Ber.* **71**: 2505 (1938); **73**: 609 (1940); C. C. Barker, E. L. Hirst, and G. T. Young: *Nature* **147**: 296 (1941). |

| | |
|---|---|
| Name | Arabic gum |
| Classification• network | Natural, organic, polysaccharides |
| Classification• crosslink point | Physical gel |
| Classification• fluid | Hydrogel |
| Manufacturing method | Extracted by scarring the bark of an acacia tree found in the Leguminosa (this type of tree grows in the wild in tropical regions, especially in the western regions of Africa in Sene-Gambia and Kordfan). The composition consists mainly of arabinic acid (79.5–81%), but this differs slightly with quality. Arabinic acid is formed by partial hydrolysis and is refined by electrodialysis. |
| Solvent | Dissolves in water. Does not dissolve in ethanol, chloroform, ethyl ether and fats and oils. |
| Characteristics | When the molecular weight is 25,000–30,000, lemon-colored masses are made, or white powders in the shape of balls with a diameter of 5–40 mm form thin cracks. The most notable characteristic is the good solubility with water. It is possible to make a highly concentrated solution of over 50%. The solubility of arabic gum in water is very high, but the viscosity is low; 40% of solution shows the property of Newtonian flow. This viscosity is affected by the pH. Normally, the pH of arabic gum water-soluble solution is 4.5–5.5, but pH 5–7 shows the highest viscosity, and when pH is 10–14 it decreases. |
| Gel preparation method | Forms a consistent solution when dissolved in water, but (when highly concentrated), will not gel. When solution is dried, forms a film. |
| Uses | Food, formulation, cosmetics, adhesives, paints, ink, printing |
| Raw material manufacturer | Sanei Yakuhin Boeki and others (import agents), Tomen, Arakawa Chemical (handling agent) |
| Product manufacturer | |
| Special note | By adding electrolyte, the normal viscosity decreases. As the concentration of electrolyte becomes high and the concentration is the same, the higher the valence of cation, the lower the viscosity. As the viscosity decreases, the sulfate tension decreases as well. It is preferred to be used as emulsifying agent. |
| Related literature | 1) Oakley: *Trans. Faraday Soc.* **31**: 136 (1935). |
| | 2) F. Smith: *J. Chem. Soc.* 274 (1939). |
| | 3) J. Jackson and F. Smith: *J. Chem. Soc.* 74 (1940). |

| | |
|---|---|
| Name | Arabinan |
| Classification• network | Natural, organic, polysaccharides |
| Classification• crosslink point | |

Classification• fluid

Manufacturing method | Add heat to a 2.5 wt% $Ca(OH)_2$ solution after adding the filtrate that was extracted and removed by water from the scroll of the beet. Add the filtrate removed from this to acetic acid and remove the pectin that precipitated. After acetylating the refined arabinan with pyridine and acetic anhydride, deacetylate and refine.

Solvent | Water/ethanol = 3/7

Characteristics | $[\alpha]_D = 114-160°$
$\bar{M}_w = 7 \times 10^3$
There are many branchings.

Gel preparation method | After dissolving in hot water, it gels when cooled.

Uses

Raw material manufacturer | Megazime (Australia)

Product manufacturer

Related literature |
1) E. L. Hirst and J. K. N. Jones: *J. Chem. Soc.* **1948**: 2311 (1948).
2) G. O. Aspinall: *The Carbohydrates, Chemistry and Biochemistry*, 2nd ed., vols. IIA, IIB, New York: Academic Press (1970).

Chemical formula

$$\rightarrow 5\text{Ara}fa1 \rightarrow 5\text{Ara}fa1 \rightarrow 5\text{Ara}fa1 \rightarrow 5\text{Ara}fa1 \rightarrow 5\text{Ara}fa1 \rightarrow$$
$$\begin{array}{ccccc} & 3 & & 3 & & 3 \\ & \uparrow & & \uparrow & & \uparrow \\ & \text{Ara}fa1 & & \text{Ara}fa1 & & \text{Ara}fa1 \end{array}$$

or

$$\left[ \begin{array}{c} \text{5Ara}fa1 \rightarrow \text{5Ara}fa1 \\ 3 \\ \uparrow \\ \text{Ara}fa1 \end{array} \right]_n$$

Ara*fa*1 = L-Arabinofuranose residue

---

Name | Sodium alginate

Classification• network | Natural, organic, polysaccharides

Classification• crosslink point | Physical gel

Classification• fluid | Hydrogel

Manufacturing method | After soaking brown algae (macro *cysteis, lessonia*) in dilute acid, heat while it is alkaline, and extract and separate as soluble salt.

Solvent | Water

Characteristics | Forms a gel that has strong heat resistance under the existence of bivalent ion. It is not always necessary to add heat to make it gel. The gel property differs according to the ratio of the mannaaric acid (M) and the gluronic acid (G). When M is rich, it is a soft, elastic gel, when G is rich the gel is strong but brittle. In the same way as pectin, gel is formed by the egg-box model, i.e. coordination bonded by bivalent cations.

Gel preparation method | After processing the solution, gel can be obtained by adding a bivalent ion. It is possible to make gels under various conditions by selecting the solubility of the salt, pH and ion enclosing.

Uses | Food thickeners, stabilizers and gelling agents. Used in jellies and frozen candy.
Used in textile industry.

Raw material manufacturer | Neutrasweet Keruko, Kimitsu Chemical, Inc., Kibun Food Chemifa

Product manufacturer | Dainippon Pharmaceuticals

Related literature      1) E. R. Morris *et al.*: *Int. J. Biol. Macromol.* **2**: 327 (1980).

Chemical formula

---

| | |
|---|---|
| Name | Propylene glycol alginate |
| Classification• network | Natural, organic, polysaccharides (modified process) |
| Classification• crosslink point | Physical gel |
| Classification• fluid | Hydrogel |
| Manufacturing method | Propylene oxide is addition bonded under pressurized heat to the carboxyl group of alginic acid. |
| Solvent | Easily soluble under cold water, hot water, acidic solvents when it is a powder. |
| Characteristics | Does not settle from calcium or metallic sodium. Shows a high viscosity (a characteristic of alginic acid) under low concentration. |
| Gel preparation method | When esterification level is 45–80%, it gels with multicharged ions such as calcium ion. |
| | When esterification level is above 80%, it does not completely gel even for multicharged ions with large atomic numbers. |
| Uses | Stabilizers for French dressing, meringue, beer |
| Raw material manufacturer | Kimitsu Chemical, Inc., Kibun Food Chemical |
| Product manufacturer | |
| Related literature | 1) Chemical Dictionary Editing Association (ed.): *Chemical Dictionary* 1, Kyoritsu Publ. (1960). |
| | 2) Chemical Industry Report, Co: 10188 Chemical Products, Chemical Industry Report, Co. (1988). |

Chemical formula

---

| | |
|---|---|
| Name | Isolichenan |
| Classification• network | Natural, organic, polysaccharides |
| Classification• crosslink point | Physical gel |
| Classification• fluid | |
| Manufacturing method | Obtained from separation precipitation by acetone as copper complex from the residual liquid when lichenan was precipitated. |

| | |
|---|---|
| Solvent | Hot water |
| Characteristics | Dissolves in cold water. |
| | Dextrorotatory $[\alpha]_D = +225°$. Colored by iodine ($\lambda_{max}600$ nm). |
| | Decomposed by $\alpha$-amylase, but does not receive the interaction of |
| | $\beta$-amylase. |
| | pp = 42–44. |
| Gel preparation method | Gels when cooled after dissolving in hot water. |
| Uses | |
| Raw material manufacturer | |
| Product manufacturer | |
| Related literature | 1) K. Abu and N. Seno: *Basics of Sugar Chemistry*, Kodansha |
| | (1984). |
| Chemical formula | |

---

| | |
|---|---|
| Name | Insulin |
| Classification• network | Natural, organic, protein |
| Classification• crosslink point | Physical gel |
| Classification• fluid | Hydrogel |
| Manufacturing method | Extract a frozen pancreas by processing with a diluted alcohol of hydrochloric acid. Add sodium chloride to the extracted solution and salt precipitate. |
| Solvent | Diluted acid and diluted alkali solution |
| Characteristics | Shows a molecular weight of 12,000, separates most when pH is 2–3, dissociates reversibly when pH is over 7.5 or under 4.0. Molecular weight is around 48,000 in the solution of pH 7.0–7.5 and when isoelectric point is pH 5.3–5.4. |
| Gel preparation method | Heat at 100°C for 30 min at pH 2 |
| Uses | Treatment of diabetes, shock treatment for schizophrenia, treatment for obesity, anorexia and malnutrition. |
| Raw material manufacturer | |
| Product manufacturer | Wako Pure Chemical Industries, other manufacturers |
| Related literature | |
| Chemical formula | The structure of cow insulin is as follows: The position of the -S-S-crosslinking between chain A and chain B is as follows: |

```
                    S ——————— S
                    |         |
    A chain —— 6 ⁻ 7 ⁻⁻ 11 ——— 20
                    |         |
                    S         S
                    |         |
                    S         S
                    |         |
    B chain ——————— 7 ——————— 19 ——
```

---

| | |
|---|---|
| Name | Ethylcellulose, cellulose ethyl ether [EC] |
| Classification• network | Natural, organic, polysaccharides |
| Classification• crosslink point | |

| | |
|---|---|
| Classification• fluid | Hydrogel |
| Manufacturing method | Wash with hot water and dry after alkali cellulose (pulp, linter pulp as original material) is reacted heterogeneously with diethyl sulfate or ethyl chloride under pressure. |
| Solvent | Alkali solvent, water, many organic solvents (see characteristics) |
| Characteristics | Linear chain polymer has no taste or odor, is a white powder and the specific gravity is 1.14. It is chemically stable and hard to burn, and is not affected by sunlight or water. It has high thermal stability and is almost not affected until it reaches the softening point (152–162°C). The solvent changes according to the increase in content of the ethoxy group from sodium hydroxide to water to a mixed solvent of toluene and ethanol to alcohol to benzene and toluene. |
| Gel preparation method | Generally, crosslinking can be done using cellulose crosslinking agents (formaldehyde, dialdehyde, diisocyanate, halohydrin). There are some that polymerize with vinyl monomers, and crosslinking agent (methylene bis acrylamide, ethylene glycol, diacrylate). |
| Uses | When the ethoxy group content is 45.5–49%, it is used as a plastic. Helmets and tools are made from this. There is a compatibility with processed rosin and drying oils. The properties that make the solvent desorption good make this a good product to mix rotogravure ink, flexo graphic ink, print-dyeing emulsion inks with other resins. |
| Raw material manufacturer | Imported = Nissho Iwai [Manufacturing source: (U.S.A.)] |
| Product manufacturer | |
| Special note | Among cellulose-type plastics, it has exceptional alkali resistance and machine strength under low temperatures, is weatherproof and stable under heat. |
| Related literature | |
| Chemical formula | |

$$\left[ \begin{array}{c} \text{CH}_2\text{OR} \\ \text{H} \overset{\text{O}}{\underset{\text{OR}}{\diagup}} \text{O}- \\ \overset{|}{\underset{\text{H}}{\diagup}} \overset{\text{H}}{\underset{\text{OR}}{\diagup}} \\ \text{H} \end{array} \right]_n \qquad \text{R}: -\text{C}_2\text{H}_5 \text{ or H}$$

| | |
|---|---|
| Name | Ethylhydroxyethyl cellulose [EHEC] |
| Classification• network | Natural, organic, polysaccharides |
| Classification• crosslink point | |
| Classification• fluid | Hydrogel |
| Manufacturing method | Make by interacting ethyl chloride and ethylene oxide with alkali cellulose. |
| Solvent | Water, aliphatic hydrocarbon |
| Characteristics | A linear chain polymer has no taste or odor, is a white powder, the specific gravity is 1.12, is soluble among a wide variety of solutions, and has the unique characteristic of dissolving in aliphatic hydrocarbons that are low in price and have little odor. |
| Gel preparation method | Unknown |
| Uses | Printing inks (silkscreen ink, rotogravure ink), clear lacquer (for plastics and coating for porous papers), other. |

| | |
|---|---|
| Raw material manufacturer | Imported = Nissho Iwai [Manufacturing source: (U.S.A.)] |
| Product manufacturer | |
| Special note | Different from other cellulose derivatives. It has the characteristic of dissolving in aliphatic hydrocarbons. There is the advantage of being able to use cheaper solutions. |
| Related literature | |
| Chemical formula | |

X: $(CH_2CH_2O)_x$ or $-C_2H_5$ or -H
Y: $(CH_2CH_2O)_y$ or $-C_2H_5$ or -H
Z: $(CH_2CH_2O)_z$ or $-C_2H_5$ or -H

---

| | |
|---|---|
| Name | Curdlan |
| Classification• network | Natural, organic, polysaccharides |
| Classification• crosslink point | Physical gel |
| Classification• fluid | Hydrogel |
| Manufacturing method | Obtained by separating the polysaccharides that yield microorganisms *Agrobacterium*, *Alcaligenes faecalis*. |
| Solvent | Water (alkali, hot water) |
| Characteristics | Linear chain polymer does not dissolve in water, but when heated to above $80°C$, it forms a strong, nonreversible gel. After it is cooled to $50–60°C$, it turns into a reversible gel. The gel heated to over $80°C$ will remain a nondissolving gel even when heated to $130°C$, and it becomes a strong gel highly heat resistant. There is considerable evaporable water. $[\alpha]_D = +18°$ (0.1 N NaOH) |
| Gel preparation method | Gels after heating the dispersed polymer in water. When the alkali solution that has mixed with other viscoelastic polysaccharides (i.e., seaweed polysaccharides such as carrageenan, vegetable seed gum such as guar gum and microorganism polysaccharides such as gellan gum) is neutralized, it gels. |
| Uses | Food gelling agents. Meat products, water-glossing, noodles, and side dishes. |
| Raw material manufacturer | Takeda Pharmaceutical Co. |
| Product manufacturer | Takeda Pharmaceutical Co. |
| Related literature | 1) A. Harada: *Fast Food Industry* **29**(3): 79 (1987). |
| Chemical formula | |

---

| | |
|---|---|
| Name | Casein |
| Classification• network | Natural, organic, protein |
| Classification• crosslink point | Physical gel |
| Classification• fluid | Hydrogel |

| | |
|---|---|
| Manufacturing method | Casein is a protein taken from milk and contains about 3% of colloids as calcium salt. Casein is obtained by adding to milk an acid that removes the Ca, coagulates it, leaving only the casein. There are four manufacturing methods of casein available on the market today: 1) The natural fermentation method; 2) coagulation through adding sulfur; 3) coagulation through adding hydrochloric acid; and 4) the rennet method. For the natural fermentation method, the extracted milk fat that is obtained from creating butter is heated to 43°C and the lactic acid is fermented. The acidity is adjusted to 0.46–0.6%, and the pH is around 4.6. It is heated to 52°C while being agitated. For 2–4 h the coagulated protein (curds) are separated and settled, then washed with water, dried and crushed. |
| Solvent | Does not dissolve in water or organic solvents. Disperses and dissolves in water when salt or alkali is added. |
| Characteristics | There are over 30 types of casein confirmed when the subclass and hereditary mutations are added. Generally, it can be divided into $\alpha$-, $\beta$-, $\kappa$-, $\gamma$-casein. The product on the market is cream- or lemon-colored, but the pure form is white. The molecular weight is 13,000–25,000, and the isoelectric point is pH 4.6. For casein, the hydrophilicity is smallest at the isoelectric point and has the properties of having minimal bonding strength with ions. Using these properties, the coagulated casein is made. |
| Gel preparation method | Gels when calcium is added and heated. |
| | Highly concentrated solutions for casein sodium will gel. |
| Uses | Mostly used for food or medicines, but it is also used in cosmetics as emulsifiers, emulsification aides or thickeners mainly as creams or latex. Because there are problems with stable quality and properties, while widely used in industrial settings, it is not used as much for cosmetics. |
| Raw material manufacturer | (Handling agent) Arakawa Chemical, Inc., Sanei Yakuhin Boeki |
| Product manufacturer | |
| Related literature | 1) K. Yamauchi and K. Yokoyama (eds.): *Milk Dictionary*, Asakura Publ. |
| Chemical formula | |

| | |
|---|---|
| Name | Carrageenan |
| Classification• network | Natural, organic, polysaccharides |
| Classification• crosslink point | Physical gel |
| Classification• fluid | Hydrogel |
| Manufacturing method | Extracted from red seaweed and *Gigaritina tenella* Harvey, using water. |
| Solvent | Water |
| Characteristics | Dissolves in hot water (above 70°C), but will not dissolve in cold water. |
| | Average molecular weight is above 100,000. |
| | Dissolves in ethanol and glycerin when concentration is around 40%, and difficult to dissolve in concentrations above that. |
| Gel preparation method | After boiling in 30% water, it is cooled and gelled. |

## 86  *Data Summary*

| | |
|---|---|
| Uses | Various food products, medicines, cosmetics and toothpaste that use the gelling, reactivity with the protein, viscosity, dispersion and suspension. |
| Raw material manufacturer | FMC, SBI, Copenhagen Pectin, MRC, Polysaccharide |
| Product manufacturer | Mitsui & Co., Sansho, Yukijirushi Foods Co. |
| Special note | There are three types of carrageenan: $\kappa$, $\lambda$, $\iota$. Those that gel with ions are the $\kappa$ and $\iota$ types. The $\iota$ type produces the hardest gel with the salt of $Ca^{2+}$ ($K^+$ for $\kappa$). |
| Related literature | 1) D. R. Morris *et al.*: *Int. J. Biol. Macromol.* **2**: 327 (1980). |
| Chemical formula | |

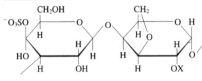

| | |
|---|---|
| Name | Carboxymethyl cellulose |
| Classification• network | Natural, organic, polysaccharides (modified process) |
| Classification• crosslink point | Physical gel |
| Classification• fluid | Hydrogel |
| Manufacturing method | Mix sulfurous acid pulp with sodium hydroxide and make an alkali cellulose, dissolve sodium chloroacetic acid, knead and precipitate it with methanol. $R_{cell}$-ONa + ClCH$_2$COOH $\rightarrow$ R$_{cell}$-OCH$_2$COOH + NaCl |
| Solvent | Swells and dissolves in basic water where the degree of derivatization of the carboxymethyl group is over 0.3. When it is over 0.4, it dissolves and swells in water. |
| Characteristics | The normal chain polymer is obtained in the form of sodium salt that is a normal hygroscopic white solid that absorbs humidity. The properties change according to the degree of the derivatization of the carboxymethyl group. When the degree of derivatization is over 0.3, it is acidic and it precipitates, but those that are acidic and are between 0.5–0.8 will not precipitate, but will precipitate in alcohol. |
| Gel preparation method | Forms a hydrogel without using crosslinking agents. After dispersing in organic solvents (glycerin and alcohols), it gels when placed into a metallic sodium solution (such as calcium hydroxide, aluminum sulfate, ferric chloride). Forms nonsoluble material by heat or crosslinking agent for cellulose-like epichlorohydrin. Free radical polymerization with divinyl monomer. Grafting or crosslinking by radiation. |
| Uses | Used as a stabilizer for textile pastes, dyeing pastes, emulsified paints, adhesives, ice cream and jams that use the characteristics of hydrophilicity, high viscosity of solutions, film formation, nonpoisonous property, protection colloids and adhesion. |
| Raw material manufacturer | Daiichi Kogyo Industries, Daicel Chemical Industries, Ltd. |

| Product manufacturer | Serva Feinbiochemica GmbH & Co. KG |
|---|---|
| Special note | Makes various types of salt using polymer electrolyte. |
| Related literature | 1) H. Tomida, C. Nakamura, and S. Kiryu: *Chem. Pharm. Bull.* **42**: 979 (1994). |
| Chemical formula | |

---

| Name | Carboxymethyl starch |
|---|---|
| Classification• network | Natural, organic, polysaccharides |
| Classification• crosslink point | Physical gel |
| Classification• fluid | Hydrogel |
| Manufacturing method | Formed when starch is reacted with monochloro acetic acid in a hydrophilic solution such as sodium hydroxide. The temperature is maintained at 20–80°C and the reaction will take from several hours up to 2 days. One reaction gives a substitution rate of around 0.5% and when a higher substitution rate is needed, reactions should be repeated. |
| Solvent | Water |
| Characteristics | It will start swelling in cold water when the substitution rate gets close to 0.15. The solution is consistent and exhibits a structural viscosity which becomes higher as the replacement rate increases. More stable when alkaline than neutral.<br>When metallic salt and strong acids exist, it will precipitate and show cloudiness. |
| Gel preparation method | Mix carboxymethyl starch powder with water and add into a calcium chloride that has a processed concentration. This creates a gel with a constant shape. By changing the concentration of the various solutions, the strength can be adjusted. |
| Uses | Collapsor for medicinal pills. Printing paste. Packing material for column chromatography. |
| Raw material manufacturer | Not available on the market. |
| Product manufacturer | |
| Special note | It has unique application due to low cost because it dissolves faster with degree of derivatization than carboxymethyl cellulose |
| Related literature | 1) H. Suzuki, Y. Tadokoro, and N. Taketomi: *Precipitation Science* **9**: 33 (1961).<br>2) H. Suzuki, Y. Hashimoto, and R. Takahashi: *Precipitation Science* **17**: 293 (1969). |
| Chemical formula | Starch -OCH$_2$COOH |

---

| Name | Callose |
|---|---|
| Classification• network | Natural, organic, polysaccharides |
| Classification• crosslink point | Physical gel |
| Classification• fluid | Hydrogel |
| Manufacturing method | Refined from the hot-water extracted part of the cell wall of sweet corn buds. |
| Solvent | Basic solution |
| Characteristics | Dyed with aniline blue and resorcinol blue and gives out a green fluorescence. |

Shows an individual protective interaction among plants when there is damage to the composition or if there are pathogenic microorganisms deposited on the cell wall and the sieve area during infection.

| | |
|---|---|
| Gel preparation method | Unknown |
| Uses | |
| Raw material manufacturer | |
| Product manufacturer | Sigma |
| Related literature | 1) Z. Hassid: The *Carbohydrates, Chemistry and Biochemistry*, 2nd ed., vols. IIA and IIB, New York: Academic Press (1970), p. 301. |
| | 2) Y. Kato and D. J. Nevins: *Plant Physiol.* **78**: 20 (1985). |
| Chemical formula | |

| | |
|---|---|
| Name | Agar |
| Classification• network | Natural, organic, polysaccharides |
| Classification• crosslink point | Physical gel |
| Classification• fluid | Hydrogel |
| Manufacturing method | Extraction from seaweed (*Gelidium amansii, ogonori, obakusa,* etc.) |
| Solvent | Water (necessary to heat to around 90°C to dissolve) |
| Characteristics | Forms a gel that is brittle and artificial. One of the most general gelation agents for food. The melting point of the gel is high but there is a tendency to lose water. |
| | There is little effect on the gelation from other factors (ions, polysaccharides). |
| | Agar is constructed from agarose and agaropectin, but agarose is a double helix structure and has a threefold axis going counterclockwise. |
| Gel preparation method | It gels after being dispersed in water, heated and then cooled. |
| Uses | Food gelling agent. Boiled bean cubes, gelidium jelly, jellies, Japanese candies. |
| | Culture medium, electrophoretic gel. |
| Raw material manufacturer | Ina Foods Industry |
| Product manufacturer | Ina Foods Industry |
| Related literature | 1) S. Arnott *et al.*: *J. Mol. Biol.* **90**: 269 (1974). |
| Chemical formula | |

| | |
|---|---|
| Name | Xanthan gum |
| Classification• network | Natural, organic, polysaccharides |
| Classification• crosslink point | Physical gel |
| Classification• fluid | Hydrogel |
| Manufacturing method | Obtained by separating the polysaccharides released from microorganism *Xanthomonas campestris*. |

| | |
|---|---|
| Solvent | Water |
| Characteristics | Linear chain polymer has low concentration and high viscosity and forms an exceptionally stable solution that is a typical pseudplastic. It will not gel only with xanthan gum, but by combining with galactomannan such as locust bean gums, a gel with a rather strong elasticity is formed. The gel strength changes depending on the composition. Helix of xanthan and the galactomannan chain with few branches associate and crosslink, leading to gelation. |
| Gel preparation method | Gels when the mixed solution of xanthan and locusts is cooled after it has been thermally processed. |
| Uses | Food thickener, stabilizer. Dressings, pickled foods, sauces. Cosmetics, toiletries, drilling for oil wells. |
| Raw material manufacturer | |
| Product manufacturer | Dainippon Pharmaceuticals, Saneigen FFI, Yukijirushi Foods Co. |
| Related literature | 1) I. C. M. Dea *et al.*: *J. Mol. Biol.* **68**: 153 (1972). |
| Chemical formula | |

$M^- = Na, K, \frac{1}{2}Ca$

---

| | |
|---|---|
| Name | Chitin |
| Classification• network | Natural, organic, polysaccharides |
| Classification• crosslink point | Physical gel |
| Classification• fluid | Hydrogel |
| Manufacturing method | In the external skeleton of crustaceans, there is not only chitin but also other calcium carbonates such as inorganic salts, proteins, fats and pigment included. These coexistent materials are removed using the following three steps: 1) Decarbonization (removal of inorganic salts) soaked in a diluted hydrochloric acid or ethylene diamine tetraacetate; 2) removing protein heated in sodium hydroxide and interacted with a protein decomposing fungus; and 3) removal of fats or pigment...treated with organic solvents. |

| | |
|---|---|
| Solvent | Concentrated hydrochloric acid, concentrated sulfuric acid (does not dissolve in water, organic solvents, weak acids, weak salt groups). |
| Characteristics | Linear chain polymers exist as a polysaccharide protein in the organism. Specific rotation $[\alpha]_D = -15°$ (in concentrated hydrochloric acid). White powder. When completely hydrolyzed in acid, it obtains a D-glucosamine acetic acid. When in alkali, decomposes to chitosan and acetic acid. Disassociates by enzyme chitinase of snails and worms or lysozyme and creates an N-acetyl-D-glucosamine or its oligosaccharides. |
| Gel preparation method | The micropowder of chitin is dispersed at room temperature using $CaCl_2 \cdot 2H_2O$ saturated methanol solutions. When this is refluxed for several hours, a consistent and clear dope is obtained. After being degassed, the film is immersed in a 20% mixed sodium formic acid aqueous solution and ethylene glycol (1 : 1 v/v) overnight, resulting in a chitin gel after calcium is removed. |
| Uses | Used for treating external injuries such as burns as a wound dressing. |
| Raw material manufacturer | Yakitsu Marine Biology Co., Katokichi |
| Product manufacturer | Biochemical Co. |
| Related literature | 1) R. Ogawa and S. Tokura: *Carbohydr. Polym.* **19**: 171–178 (1992). |
| | 2) A. G. Walton and J. Blackwell: *Biopolymers* 474–489 (1973). |

Chemical formula

| | |
|---|---|
| Name | Chitosan |
| Classification• network | Natural, organic, polysaccharides |
| Classification• crosslink point | Physical gel |
| Classification• fluid | Hydrogel |
| Manufacturing method | Obtained due to the docetylation of the N-acryl group by heating chitin in a 40% highly concentrated alkali solution or heating powdered chitin in fused calcium hydroxide at 180°C for 30 min. |
| Solvent | Diluted acid (not soluble with water, ethanol, and others) |
| Characteristics | The viscosity range for linear chain polymer: 800–1300 cPs (chitosan 1000), a noncrystallized powder with no color. Gives a purple color by processing with iodine and sulfuric acid. |
| Gel preparation method | Gels when neutralized after dissolving in dilute acid. |
| Uses | Controlled release formation (i.e., controlled release of aspirin or indomethacin from chitosan granules). Functional separation materials (column fillers). |
| Raw material manufacturer | Katakura Chikkarin, Katokichi |
| Product manufacturer | Biochemical Co. |

Related literature

1) Chitin, Chitosan Research Association: *Chitin, Chitosan Testing Manual*, Gihohdoh (1991).
2) S. Tokura: *Cellulosics Utilization* (H. Inagaki and G. O. Phillips eds.) New York: Elsevier Applied Science (1989), p. 63.

Chemical formula

Name | Fibroin
---|---
Classification• network | Natural, organic, protein
Classification• crosslink point | Chemical gel
Classification• fluid | Hydrogel
Manufacturing method | Repeat the operation of taking the fiber of a cocoon and heating for around three hours at 120°C under pressure with around 25 times the water until the weight starts decreasing. Process this for 24 h in 1% hydrochloric acid and wash it with water, ethanol and ether (in this order).
Solvent | Solution such as lithium bromide, calcium chloride and calcium nitrate
Characteristics | Does not dissolve in water, dilute acid or dilute alkali. Is stable in protease and has some reversible elasticity. The length shrinks by swelling when processed with dilute formic acid.
Gel preparation method | Refine the liquid fibroin, which is the silkworm's silk thread material. Cut off around 3 cm from the junction of the middle area and end area in the middle thread, remove the gland enzyme, and leave it in the water. The sericin is removed after about 30 min. When the sericin is washed off, a liquid fibroin is left as gel-like.
Uses | Used as fibers
Raw material manufacturer |
Product manufacturer | Wako Pure Chemical Industries
Related literature |
Chemical formula |

| | |
|---|---|
| Name | Guar gum |
| Classification• network | Natural, organic, polysaccharides |
| Classification• crosslink point | Physical gel |
| Classification• fluid | Hydrogel |
| Manufacturing method | Obtained by crushing the fine powder that was taken out of the endosperm area by crushing the embryo of the seed of *Cyamopsis tetragonolobus*. After cooling this, remove the precipitate so that the ethanol concentration will be 25 wt% and further by adding ethanol to make it 40% for final recovery. |
| Solvent | Water |
| Characteristics | $[\alpha]_D = +53°$ |
| | Colloidal dispersion fluid shows thixotropy. |
| | Viscosity of 1% aqueous solution is 3000–6000 cP. |
| | Dissolves well in cold water. Viscosity of sol does not change in the range of pH = 1–10.5. |
| Gel preparation method | Gels by adding $Ca^{2+}$, $Al^{3+}$, $Cr^{3+}$, $Na_2[B_4O_5(OH)_4] \cdot 8H_2O$ |
| Uses | Food additives (cheese, ice cream, arum root). |
| | Adding paste to weaving, drilling for oil wells. |
| Raw material manufacturer | Mayhall, Pakistan Gum & Chemical, SBI |
| Product manufacturer | Rhone Poulenc, Dainippon Pharmaceuticals, Sansho, Yukijirushi Foods |
| Related literature | 1) E. Heyne and R. L. Whistler: *J. Am. Chem. Soc.* **70**: 2249 (1948). |
| | 2) R. L. Whistler and D. J. Durso: *J. Am. Chem. Soc.* **73**: 4189 (1951). |
| Chemical formula | |

| | |
|---|---|
| Name | Quinc seed gum |
| Classification• network | Natural, organic, polysaccharides |
| Classification• crosslink point | Physical gel |
| Classification• fluid | Hydrogel |
| Manufacturing method | Extracted from the viscous liquid in the seed of a *Cydonia oblonga* tree that grows in Europe and Asia using hot water. When removing the seed, the fruit is rotted and air-dried, then the pulp, dirt and other impurities are removed and the seed is removed. Normally, the seed is extracted using water for 30 min with 100 parts of cold or hot water to every two parts of seed. The mucilage is removed by natural filtration using a muslin cloth. |

| | |
|---|---|
| Solvent | Water |
| Characteristics | The production amount of seeds from *Cydonia oblonga* trees is very little. The extracted mucilage has a unique texture (smooth and not sticky) compared to other natural mucilage and synthetic polymers. In the mucilage after extraction, there is 33% of cellulose, and as a component sugar, L-arabinose, D-xylose, glucose, galactose, hexronic acid or monomethyl hexronic acid exists. |
| Gel preparation method | After heating and dissolving 1% solution, when cooled, a weak gel is formed. |
| Uses | Used for hand lotions or setting lotions that use latex or glycerin. |
| Raw material manufacturer | |
| Product manufacturer | |
| Special note | The disadvantage of this product is the extremely weak resistance to microorganisms. When using it, it is necessary to either disinfect the mucilage or add germ-preventing agents. |
| Related literature | 1) A. G. Renfrew and L. H. Cretcher: *J. Biol. Chem.* **97**: 50 (1932). |
| Chemical formula | |

---

| | |
|---|---|
| Name | Crown gall |
| Classification• network | Natural, organic, polysaccharides |
| Classification• crosslink point | Physical gel |
| Classification• fluid | Hydrogel |
| Manufacturing method | Culture *agrobacterium tumefaciens* in sucrose, glucose, fructose for five days at 26°C and purify it by precipitating in ethanol. |
| Solvent | Hot water |
| Characteristics | |
| Gel preparation method | Cool after dissolving in hot water. ($[\alpha]_D = -9°$ or $-10°$, $M_w = 3600 \pm 200$.) |
| Uses | |
| Raw material manufacturer | |
| Product manufacturer | |
| Related literature | 1) F. C. McIntire, W. H. Peterson, and A. J. Riker: *J. Biol. Chem.* **143**: 491 (1942). |
| | 2) E. W. Putnam, A. L. Potter, R. Hodgson, and W. Z. Hassid: *J. Am. Chem. Soc.* **72**: 5024 (1950). |
| Chemical formula | |

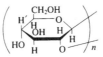

---

| | |
|---|---|
| Name | Glycogen |
| Classification• network | Natural, organic, polysaccharides |
| Classification• crosslink point | Physical gel |
| Classification• fluid | |

| | |
|---|---|
| Manufacturing method | Glycine buffer (0.2 mol. pH 10.5) that is four times the weight of the animal composition (liver, muscle) is ground in a Waring blender with twice the chloroform (washed with water). Separate the water layer using centrifugation, and add twice the glycine buffer to the middle layer and chloroform layer to reprocess. Repeat this operation five times, combine the water layers obtained and freeze at $-20°C$. Defrost the freeze-protected material and remove the nondissolved particles using centrifugation. Glycogen is obtained as a precipitate when the strained material is centrifuged. |
| Solvent | Hot water, water, formaldehyde, dimethyl sulfate |
| Characteristics | $[\alpha]_D = +191-199°$ (water) Molecular weight $10^7$ Gives a reddish-brown color with the iodine-starch reaction. Acid causes hydrolysis; is stable in alkali. |
| Gel preparation method | Unknown |
| Uses | |
| Raw material manufacturer | |
| Product manufacturer | Nakarai Tesque |
| Related literature | 1) C. Bernard: *Leçons sur le diabete*, p. 553, Paris (1877). 2) E. Bueding and S. A. Orell: *J. Biol. Chem.* **239**: 4018 (1964). |
| Chemical formula | |

| | |
|---|---|
| Name | Glucomannan |
| Classification• network | Natural, organic, polysaccharides |
| Classification• crosslink point | Physical gel |
| Classification• fluid | Hydrogel |
| Manufacturing method | Obtained by drying, powdering or alcohol refining the arum root. |
| Solvent | Water |
| Characteristics | The linear chain polymer forms extremely high-viscosity aqueous solution. Forms a gel that is elastic and thermally nonreversible under existence of alkali. The gel strength multiplies with the simultaneous use of xanthan gum and carrageenan. The interaction between arum root mannan and xanthan is stronger than the interaction between xanthan and galactomannan. |
| Gel preparation method | Gel is formed after dissolving in water, adding alkali such as calcium hydroxide and cooling after it has been heated. |
| Uses | Food gelling agents. Jellies and noodles |
| Raw material manufacturer | Shimizu Chemical |
| Product manufacturer | Shimizu Chemical |
| Related literature | 1) I. C. Dea *et al.*: *Carbohydr. Res.* **57**: 249 (1977). |

Chemical formula

| Name | Keratan sulfate |
|---|---|
| Classification• network | Natural, organic, polysaccharides |
| Classification• crosslink point | |
| Classification• fluid | Hydrogel |
| Manufacturing method | Dissolve in acetic acid+calcium acetate solution the entire glycosaminoglycan, which is precipitated by adding salt and ethanol after digesting the test material with a protein-digesting enzyme. While cooling this by ice water, add ethanol and make it precipitate with the concentration of ethanol at 45–65 wt%. Refine using CPC fractionation, gel filtration, zone migration, negative ion exchange chromatography and fractional elution. |
| Solvent | Water |
| Characteristics | $[\alpha]_D = +4.5°--10°$ $M_w = 1-2 \times 10^4$ |
| | Precipitates in 50–65% ethanol solution. Hydrolysis occurs easily by acid to compositional monosaccharide units. The amount in the human costal cartilage increases with age. |
| Gel preparation method | Unknown |
| Uses | |
| Raw material manufacturer | Sigma |
| Product manufacturer | |
| Related literature | 1) K. Meyer, A. Linker, E. A. Davidson, and B. Weissman: *J. Biol. Chem.* **205**: 611 (1953). |
| | 2) M. D. Comper and T. C. Laurent: *Physiol. Rev.* **58**: 255 (1978). |

Chemical formula

| Name | Keratin protein |
|---|---|
| Classification• network | Natural, organic, protein |
| Classification• crosslink point | Chemical gel |
| Classification• fluid | Hydrogel |
| Manufacturing method | Make powder out of hair, and digest and eliminate the various proteins using appropriate protease such as pepsin after washing with hot organic solvents and then hot water. |

Solvent                                      TGA solution
Characteristics                              Rich in cystine residues (SH group) with average molecular
                                             weight around 50 kDa.
                                             Made from HS and LS proteins.
Gel preparation method                       After hairs were reduction processed for 2 to 300 min using 0.5 M
                                             thioglycol acid solution, if the SH group is blocked using
                                             N-ethylmaleimide, then at 8 MliBr/Bc dilution, the hair will show
                                             a rubberlike elasticity.
Uses                                         Test medicines for immunological research.
Raw material manufacturer                    Canadian Biochemical, Co
Product manufacturer
Related literature                           1) *Chemistry Dictionary*, Kyoritsu Publ.
                                             2) Japan Biochemistry Association (ed.): *Biochemistry Data
                                                Book*, Tokyo Kagaku Hojin.
                                             3) Japan Analytical Association (ed.): *Polymer Analysis
                                                Handbook*, Kinokuniya Shobo.
Chemical formula

| | Hard keratin (hair, nail, horn etc.) | | | Soft keratin (skin keratin layer) | |
|---|---|---|---|---|---|
| | LS protein | HS protein | Glycin-rich protein | LSA protein | HS protein |
| Composition (%) | 40–85 | 5–45 | 1–30 | ~93 | ~17 |
| Molecular weight | ~50,000 | 10,000–30,000 | 5,000–10,000 | ~60,000 | 20,000–40,000 |
| Cysteine residue (%) | 4 | 10–30 | 5–11 | 1 | 2 |
| Glycine residue (%) | 3 | 3–9 | 20–40 | 14 | 17 |
| Helical content (%) | ~50 | 0 | 0 | ~45 | – |
| Origin | microfibril | matrix | matrix | fiber | unknown |

Name                                         Florideanstarch
Classification• network                      Natural, organic, polysaccharides
Classification• crosslink point              Physical gel
Classification• fluid                        Hydrogel
Manufacturing method                         Extracted using hot water from seaweed. The galactan sulfuric
                                             acid that is an impurity is removed using the ion exchange.
Solvent                                      Water
Characteristics                              $[\alpha]_D = +176°$
                                             Gets a reddish-brown color from iodine.
                                             Decompose by R enzyme or $\beta$-amylase.
                                             The particle of Florideanstarch is smaller than normal starch
                                             powders.
Gel preparation method                       Gels when heated in solution.
Uses
Raw material manufacturer
Product manufacturer
Related literature                           1) B. J. D. Meeuse, M. Andrie, and J. A. Wood: *J. Exptl. Botany*
                                                **11**: 129 (1960).
                                             2) P. O. Colla: *Proc. Roy. Irish Acad.* **B55**: 321 (1953).
Chemical formula

| Name | Yeast mannan |
|---|---|
| Classification• network | Natural, organic, polysaccharides |
| Classification• crosslink point | |
| Classification• fluid | |
| Manufacturing method | Add a citric acid buffer (pH = 7) to a water-dispersion fluid of bread yeast, and sterilize at high pressure for 2 h at 140°C. Cool the reaction fluid and centrifuge, high-pressure sterilize the residual liquid again, and decrease pressure and enrich it. Include the precipitate formed by adding acetic acid and neutralizing. Then precipitate the crude mannan using ethanol. The refining is done by ethanol and Fehling's solution. |
| Solvent | Water |
| Characteristics | $[\alpha]_D = +88°$ |
| | Shows antigenicity and antitumorigenicity. |
| Gel preparation method | Unknown |
| Uses | |
| Raw material manufacturer | Nakarai Tesque |
| Product manufacturer | |
| Related literature | 1) S. Peat, W. J. Whelan, and T. E. Edwards: *J. Chem. Soc.* **1961**: 29 (1961). |
| Chemical formula | |

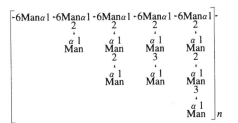

| Name | Collagen |
|---|---|
| Classification• network | Natural, organic, protein |
| Classification• crosslink point | Physical gel |
| Classification• fluid | Hydrogel |
| Manufacturing method | (1) Extract the collagen's mixed materials using organic solvents, wash them with water and then extract using a diluted salt solution. Process using acid or alkali, remove using enzyme interaction, and leave the collagen as a nonsoluble material. (2) After degreasing fresh skin, extract using 0.06 M citric acid buffer (pH 4). After dialysis, extract using 0.5 M hydrogenphosphate-2-sodium, extract the precipitate that is created in the dialysis again using 0.2 M citric acid buffer (pH 3.8). When this is dialyzed, a recycled collagen is created. |
| Solvent | Citric acid buffer |
| Characteristics | Form five types of collagen aggregates from collagen solutions. |

Produce a collagen fiber using 0.9% NaCl. FLS is obtained by adding serum-sugar protein to collagen solutions and is created when dialyzed to water. SLS is formed when adding ATP to collagen solutions. Short-period aggregates are created when NaCl is added to collagen solution to make it 2%. Amorphous aggregates are created from 5% NaCl. Each aggregate is dissolved using dilute acids and can be reverted again into any type of aggregate.

| | |
|---|---|
| Gel preparation method | Gels when the concentration of the collagen solution is increased. When boiled for a long time in water, dilute acid or dilute alkali, it changes into a gelatin made of a protein derivative. |
| Uses | Culture group *in vivo* of various cells |
| Raw material manufacturer | |
| Product manufacturer | Kanto Chemical, Koken |
| Related literature | 1) H. Hagino, Y. Osada, T. Fushimi and A. Yamauchi (eds.), *Gel-Basics and Application of Soft Materials*, Sangyo Tosho |
| Chemical formula | One-third of the string of amino acids is Gly and are in rows of threes. |

| | |
|---|---|
| Name | Thermistor |
| Classification• network | Synthetic, inorganic |
| Classification• crosslink point | Chemical gel |
| Classification• fluid | |
| Manufacturing method | In the $Ba(OEt)_2$ solution obtained by the reaction of Ba and ethanol, add $Ti[OCH(CH_3)_2]_4$ $(Y(Obu)_3$ $(Si(OEt)_4$ solution and make it hydrolyzed. Take the obtained powder and calcinate (1070°C for 2 h), then bake it at 1320°C for 2 h under pressure. |
| Solvent | |
| Characteristics | Near the phase-transition point (around 180°C), the electrical resistance drastically increases. |
| Gel preparation method | In the $Ba(OEt)_4$ ethanol solution obtained by following $Ba+2EtOH–Ba(OEt)_2$, add and mix $Ti[OCH(CH_3)_2]_4$ ethanol solution, organic solvent, water and acid. |
| Uses | Used as PTC thermistor, excess current protection element low-temperature heat, temperature sensor, condensor. |
| Raw material manufacturer | Wako Pure Chemical Industries |
| Product manufacturer | |
| Related literature | 1) R. Ozaki, K. Kawasaki, and M. Uemura: *Sera-kyo Annual Speech Manuscripts*, p. 500 (1991). <br> 2) S. Sakuhana: *Science of Sol-Gel Methods*, Agne Shofusha (1988). |
| Chemical formula | |

| | |
|---|---|
| Name | Cellulose acetate |
| Classification• network | Natural, organic, polysaccharides (modified) |
| Classification• crosslink point | Chemical gel |
| Classification• fluid | Organogel |

| | |
|---|---|
| Manufacturing method | Acetylate cellulose by acetic anhydride, acetyl chloride and ketene under appropriate catalyst. Industrially, it is acetylated using anhydrous acetic acid and sulfuric acid. |
| Solvent | Methylene chloride: mixed solvent of ethanol = 9 : 1, ester |
| Characteristics | A completely acetylated 3-acetic acid cellulose is crystalline and takes two types of the crystal structure of acetyl cellulose, I or II. The crystal structure can be interchanged by adding heat. The hygroscopicity and specific gravity decrease with the increase in the amount of acetic acid. By heating to a high temperature or irradiating by ultraviolet light oxygen is absorbed and decays. |
| Gel preparation method | Crosslink casting liquids such as acetone-formaldehyde and acetone-perchloric acid magnesium solution and methylene chloride-ethanol by the appropriate crosslinking agents used with the polysaccharides. |
| Uses | Textiles, paints, plastics, film, chromatography carrier, humidity sensors. |
| Raw material manufacturer | Aldrich, Acros Organics, Wako Pure Chemicals Industries, Tokyo Chemical |
| Product manufacturer | Daicel Chemical Industries, Ltd., Teijin |
| Related literature | 1) J. P. Randin and F. Zullig: Relative humidity measurements using a coated piezoelectric quartz crystal sensor. *Sensors and Actuators* **11**: 319–328 (1987). |
| Chemical formula | |

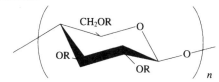

$$R = H \text{ or } COCH_3$$

| | |
|---|---|
| Name | Gellan gum |
| Classification• network | Natural, organic, polysaccharides |
| Classification• crosslink point | Physical gel |
| Classification• fluid | Hydrogel |
| Manufacturing method | Obtained by dissociating the polysaccharides released from the microorganism *Pseudomonas elodea*. |
| Solvent | Water (hot water) |
| Characteristics | Forms a heat-resistant and clear gel with low concentration (0.2%) under the existence of ions. The gel has a hard and brittle feeling when eaten. There is a tendency to lose water although it has good flavor upon release. The K-type gellan gum is found by x-ray diffraction to have a double helix structure. It is thought the association of the helices forms a 3D network structure. |
| Gel preparation method | Gels when it is heated and cooled after dispersed in water. Gelation is accelerated when cations exist. |
| Uses | Food gelling agents. Jellies, jams, Japanese candies, flour paste. |
| Raw material manufacturer | Neutrasweet Kerko |
| Product manufacturer | Dainippon Pharmaceuticals, Saneigen FFI |

| Related literature | 1) G. R. Sanderson and R. C. Clark: *Food Technology* **37**: 63 (1983). |
| | 2) K. Yamatotani: *Food Technology* **1988-9-30**: 45 (1988). |

Chemical formula

| | |
|---|---|
| Name | Schizophyllan |
| Classification• network | Natural, organic, polysaccharides |
| Classification• crosslink point | Physical gel |
| Classification• fluid | Hydrogel |
| Manufacturing method | A *S. commure fungus* is cultured for 4 to 8 days at 28°C and then the fungus is removed as precipitate after being diluted and centrifuged. It is filtered and reduced pressure-dried after it is dehydrated using 35% methanol and ether. |
| Solvent | Water |
| Characteristics | $[\alpha]_D = +2.0°$ |
| | Has a triple helical structure. |
| | It is a powder with no color or taste, and is difficult to dissolve in water; however, it slowly dissolves and becomes a liquid with high viscosity. |
| | It has antitumor activity. |
| Gel preparation method | Add sorbitol to the schizophyllan solution. |
| Uses | Medicines |
| Raw material manufacturer | Daito |
| Product manufacturer | Biochemical Co. |
| Special note | Store at −20°C. |
| Related literature | 1) S. Kikumoto *et al.*: Japan Pat. 67-12,000 (1967 to Taito), C.A.67, 107386 r (1967). |
| | 2) K. Takeo and S. Tei: *Carbohydrate Res.* **145**: 293 (1986). |
| | 3) S. Kikuchi, Miyajima, T. Yoshiseki, and S. Fujimoto: *J. Agricul. Chem., Jpn.* **44**: 337 (1970). |

Chemical formula

| | |
|---|---|
| Name | Divinylbenzene [DVB] |
| Classification• network | Synthetic, organic, vinyl polymer |
| Classification• crosslink point | Chemical gel |
| Classification• fluid | Organogel |
| Manufacturing method | After reacting the phthalaldehyde that corresponds to the structures of various isomers with Grignard's reagent, it is synthesized by dehydrogenation. Industrially, it is sometimes synthesized by dehydrogenation of diethylbenzene. Among the three types of isomers, the orthoisomer cyclizes from heating and becomes naphthalene, so generally, meta- or paraisomers are used. |
| Solvent | Dissolves in almost all aromatic solvents. |
| Characteristics | Divinylbenzene is easily radically polymerized or cationically polymerized; thus it is necessary to store or ship it in small quantities and add a large amount of polymerization inhibitor. |
| Gel preparation method | Homopolymerized or copolymerized with vinyl polymers. Forms a reactive microgel with double bonds by emulsion polymerization. |
| Uses | Homopolymers are brittle, do not dissolve or melt, and have no practical value. However, it is one of the most important crosslinking agents. In industry, they are used as ion-exchange resins, polyester resins, modified polystyrene, modified butadiene-styrene rubber. The microgel is mainly used for paints. |
| Raw material manufacturer | Sankyo Chemical |
| Product manufacturer | Kishida Chemical, Wako Pure Chemical Industries |
| Related literature | 1) Chemical Dictionary Editorial Board (ed.): *Chemical Dictionary* 4, Kyoritsu Publ. (1960). |
| Chemical formula | |

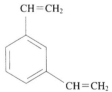

| | |
|---|---|
| Name | 2-(Dimethylamino)ethylmethacrylate [DMMA], [DAM] |
| Classification• network | Synthetic, organic, vinyl polymer |
| Classification• crosslink point | Chemical gel |
| Classification• fluid | Hydrogel, organogel |
| Manufacturing method | Ester exchange between methylmethacrylic acid and dimethylaminoethyl alcohol. |
| Solvent | Organic solvent (gradually hydrolyzed) |
| Characteristics | Melting point: approximately 30°C, boiling point: 186°C, 68.5°C (10 mmHg), specific gravity: $d_4^{20}$ 0.933, index of refraction: $n_D^{20}$ 1.4391, flashpoint: 74°C (open), viscosity: 1.38 cPs (room temperature), copolymerization constant: $Q = 0.68$, $e = 0.47$. |
| Gel preparation method | When mixed with polyacrylic acid under the appropriate concentration, solvent and ion strength, a polyelectrolyte complex gel is formed by electrostatic bonding. |

After quaternalizing the copolymer of the light crosslinkable monomer, 4′-methacryloyl oxycalcon with 1-bromo propane, a comb-shaped gold electrode is either deposited or painted onto a baked ceramic printed circuit board, and by crosslinking this with light, a moisture-sensing film can be produced.

Uses

Monomer: Adhesive by copolymerized synthetic resin, modifier for dyes, aggregating agents, static controller, ion-exchange resin, cationic comonomers for paint resins, lubricant, raw material for fuel-oil additives.

Raw material manufacturer

Wako Pure Chemical Industries, Junsei Chemical, Tokyo Chemical, Kishida Chemical, Nakara Tesque, Aldrich

Product manufacturer

Special note

Oral toxicity for mice LD50 = 1500 mg/kg.

Related literature

1) US Patent 2,832,800 (1958).
2) Functional polymer, Ch. 8, *Polymer Testing*, Vol. 7, Kyoritsu Publ. (1974).

Chemical formula

$$-(CH_2-\underset{\underset{C=O}{|}}{\overset{\overset{CH_3}{|}}{C}})n-$$

$$
\begin{array}{c}
CH_3 \\
| \\
-(CH_2-C)n- \\
| \\
C=O \\
| \\
O \\
| \\
CH_2 \\
| \\
CH_2 \\
| \\
N \\
CH_3 \quad CH_3
\end{array}
$$

---

| | |
|---|---|
| Name | Silastic Gel™ sheeting |
| Classification• network | Natural, organic, condensation polymerization |
| Classification• crosslink point | Chemical gel |
| Classification• fluid | Organogel |
| Manufacturing method | Material... made from hydrolysis of polyorganosiloxane (polydimethylsiloxane): over 99 wt%, platinum catalyst: trace monomer (dimethylsilanediol): organodichlorosilane, organodiaminosilane, organoalkoxysilane. Polymer (polydimethylsiloxane) made by condensation polymerization of dimethylsilanediol. |
| Solvent | Silicone oil |
| Characteristics | Polymer (A fluid): outer appearance: no color, clear liquid, viscosity (cP): 420 (25°C), specific gravity: 0.97. Gel (after mixing): silicone gel takes a crosslinked silicone sponge polymer to a lyogel-type that is a viscoelastic unit, swollen with the noncrosslinked solvent (oil-like silicone oligomer). Outer appearance: no color, clear. Odor: slight odor, water-solubility (g/l): nonsoluble, specific gravity: 0.97, viscosity (cP): 400 (25°C). |
| Gel preparation method | It becomes a gel in around 24 h by mixing the A fluid and B fluid in the weight ratio or volume ratio of 1 to 1 at 25°C for a working time of around 6 h. |

| | |
|---|---|
| Uses | Silastic gel sheet: Used to treat hyperplasia scars as a skin protection sheet that is not used for open wounds. |
| | Silpot 300 A&B: Used as a silicone potting material, designed for sealing and electrically protecting hybrid electrical parts for electrical circuits. |
| Raw material manufacturer | |
| Product manufacturer | Dow Corning USA (Dow Corning Asia) |
| Special note | Danger: Flammable material (flashpoint over 200°C). Stores static electricity. Hydrogen gas is created when mixing A fluid and B fluids. |
| | Safety: Dangerous: stable when following normal handling conditions. |
| | Other: Flames when mixing with acidic materials. Produces incompletely burned materials, formaldehyde and metal oxide dust. |
| Related literature | 1) RTB rubber data sheet (Dow Corning Asia) |
| Chemical formula | |

| | |
|---|---|
| Name | Silica gel |
| Classification• network | Synthetic, inorganic |
| Classification• crosslink point | Chemical gel |
| Classification• fluid | Xerogel |
| Manufacturing method | Generally, the alkali water glass is neutralized, and it is created from a dehydrated gel. The higher the degree of dehydration, the greater the ability to adsorb. |
| Solvent | Water |
| Characteristics | Absorption strength is strong for water. It is either no color or a lemon color, and is a powder of either clear or semiclear amorphous silisic acid. Water content is 2–10%, Mho hardness is 4.5–5, and density 2–2.5 $g/cm^3$. It is porous and there are some that are $450 \, m^2/g$. |
| Gel preparation method | Create solution that has composition of $Si(OC_2H_5)_4$ 169.5 g, $H_2O$ 14.7 g, $C_2H_5OH$ 324 ml. When this is hydrolyzed and polymerized a 1D long particle is created. This is made by pulling it through a nozzle when the viscosity is over 10 p. |
| Uses | Eliminating the moisture in the air; removing benzene from coal gas; removing the low boiling point hydrocarbon from natural gas; and used for chromatography filler. |
| Raw material manufacturer | Nakarai Tesque |
| Product manufacturer | |
| Special note | The filler that is modified on the silanol group by various chemical groups is used for high-performance liquid chromatography. |
| Related literature | 1) S. A. Greenberg: *J. Chem. Ed.* **36**: 218–219 (1959). |
| Chemical formula | |

| | |
|---|---|
| Name | Silicone (oil~rubber) |
| Classification• network | Natural, organic, condensation polymerization |

| | |
|---|---|
| Classification• crosslink point | Chemical gel |
| Classification• fluid | Organogel |
| Manufacturing method | Chlorosilane is obtained by reacting methyl chloride and metallic silicon under copper catalysts when the metallic silicon, produced by high-temperature reduction of the quartz, is used as a basic material. The hydrolysis of the chlorosilane gives polysiloxane oligomer but the type of chlorosilane used at this time is what gives the linear chain oligomer. Combine these polysiloxane oligomers and by polymerizing them with acid or alkali catalysts create silicone polymer. When adopting various reactive functional groups, reactive silicon oil is obtained. |
| Solvent | Silicone oil |
| Characteristics | A stable characteristic is obtained as the change in viscoelectricity is small in the temperature range of low to high temperature and the gel condition can be maintained. |
| | The polysiloxane bonding has a chemically stable structure and has exceptional solvent resistance and weatherproofing. The large stress-relaxation effect from the low elasticity rate and the high dissipation factor makes a high damping rate. By adjusting the molecular weight, the functional group and silicone oil content of the raw materials can make the adjustment of the gel characteristics possible. It is possible to add various other features by adding additives and fillers. |
| Gel preparation method | Partial crosslinking of silicone oil that has a reactive functional group. Crosslinking methods including hydrosylilation and radical polymerization. |
| Uses | Stress relaxation, vibration damping, collision buffer, packaging material adhesive (for electrical electronic use, general industrial use, sports, and medical use). |
| Raw material manufacturer | Shin-Etsu Chemical Co., Toray Industries, Inc., Dow Corning, Silicone, Toshiba Silicone, other silicone manufacturers |
| Product manufacturer | Same as Raw material manufacturer. |
| Special note | Use caution as there is the tendency for hardening inhibition to occur if the compounds that react easily with reactive functional groups and catalyst come in contact. |
| Related literature | 1) K. Ito (ed.): *Silicone Handbook*, Nikkan Kogyo Shimbunsha. |
| Chemical Formula | |

$$-\left(\,\underset{\underset{R}{\displaystyle |}}{\overset{\overset{R}{\displaystyle |}}{Si}}-O\,\right)_n-$$

$R$ : $CH_3$, $C_6H_5$, $C_nH_{2n+1}$, $CF_3CH_2CH_2\cdots$

| | |
|---|---|
| Name | Gelatin |
| Classification• network | Natural, organic, protein |
| Classification• crosslink point | Physical gel |
| Classification• fluid | Hydrogel |

| | |
|---|---|
| Manufacturing method | Refine the processed collagen formed by animal skin, bones and tendons in acid, alkali or hot water that is concentrated and dried later. Using a more moderate processing method refines a protein with a higher molecular weight. |
| Solvent | Hot water, glycerin-mixed solutions, acetic acids |
| Characteristics | Protein has a molecular weight of 100,000–250,000. Thin layers, fragments or powders that have no color or a lemon color, no flavor or smell. Does not dissolve in cold water, but it softens and swells 5 to 10 times its size when placed in water. Dissolves in warm water and becomes a consistent sol. Cooling makes it an elastic gel. It undergoes a reversible sol-gel transition, but when this is repeated several times it loses this function. |
| Gel preparation method | Place the gelatin powder in warm water (over approximately 2%) and create a clear solution. Cooling this will create a clear or semiclear gel. Increasing the concentration makes it stronger. |
| Uses | Thickeners for cosmetics. Food processing. Capsules, pills and lozenges for medicinal use. Culture area for bacterial tests. Adhesives, photograph emulsion film. |
| Raw material manufacturer | Nitta Gelatin, Nippi, Miyagi Chemical |
| Product manufacturer | |
| Related literature | |
| Chemical formula | |

$$H_2N \underset{\underset{R_1}{|}}{\overset{\overset{H}{|}}{O}} \overset{CO}{} \overset{O}{} \underset{NH}{} \overset{H_2C\ CH_2}{\overset{R_2}{N}} \underset{CO}{\overset{CH}{}} \overset{CO}{}$$

R = Amino acid

---

| | |
|---|---|
| Name | Cellulose |
| Classification• network | Natural, organic, polysaccharides |
| Classification• crosslink point | Physical gel |
| Classification• fluid | Hydrogel |
| Manufacturing method | Most of the arboreous cotton textiles are from cellulose and have few impurities. If degreased arboreus cotton is boiled under nitrogen with 1% sodium hypochloride solution, over 99% of cellulose would be extracted. Wooden cellulose mixes fine wooden powder with sodium chlorite and acetic acid, then lignin is oxidatively decomposed. From the mixed cellulose and the hemicellulose, the hemicellulose is extracted and removed by concentrated alkali. Purity is 90–97%. |
| Solvent | Dissolve in ethylene diamine, and copper ammonia solution. Does not dissolve in water. |
| Characteristics | Cellulose fibers include a different ratio of crystals and amorphous regions based on origins. Cellulose is a polymer where the D-glucose is connected with $\beta$-1,4 glucoside bonding. The degree of polymerization is from several thousand to 10,000. The length of the cellulose molecule is around 3000 nm. The dry strength is 3–5 g/d, but the wet strength can rise to 6.5 g/d. $[\alpha]_D = -120°$ (copper ammonia solution) |

| | |
|---|---|
| Gel preparation method | Dilute the dichloromethane solution of triacetic acid cellulose with heptanol and add this mixed solution to a gelatin aqueous solution. Agitate and add heat to 30°C and remove the dichloromethane that is in the suspended particle. When placing this particle in a sodium hydroxide solution, the ester is saponified and a ball-like particle of cellulose is obtained. |
| Uses | Pulp, when just gathering cellulose fibers; general clothing, towels, futons, gloves, string. Ropes for industrial use, fish nets, various nets. As sanitary products, degreasing cotton, gauze, bandages. |
| Raw material manufacturer | E. Merck Darmstadt |
| Product manufacturer | Nakarai Tesque |
| Related literature | 1) C. Doree: *The Methods of Cellulose Chemistry*, London: Chapman & Hall (1947).<br>2) P. Andrews: *The Biochemist's Handbook*, C. Long, ed., London: E. & F. N. Spon Ltd. (1961), p. 946. |
| Chemical formula | |

| | |
|---|---|
| Name | Ivory nut mannan |
| Classification• network | Natural, organic, polysaccharides |
| Classification• crosslink point | Physical gel |
| Classification• fluid | Hydrogel |
| Manufacturing method | Mannan A: Remove the wax material from the seed of the ivory nut and extract this using 7 wt% KOH solution. Neutralize this using acetic acid and precipitate the raw mannan A using ethanol. Continue refining using KOH, Fehling's solution, diluted hydrochloric acid and ethanol.<br>Mannan B: Dissolve the residual KOH extraction from mannan A in a copper ammonium solution, add NaOH and get a precipitated product. Add acetic acid to this and dissolve, and precipitate mannan B with ethanol. Refine using KOH, formic acid and ethanol. |
| Solvent | Mannan A: KOH solution, Mannan B: copper ammonium solution |
| Characteristics | Mannan A: dissolves in KOH solution, forms crystals, $[\alpha]_D = -28°$ (formic acid), the degree of polymerization = 10–13.<br>Mannan B: difficult to dissolve in KOH aqueous solution, microfibril structure, $[\alpha]_D = -20°$ (formic acid), the degree of polymerization = 38–40. |
| Gel preparation method | Unknown |
| Uses | Adhesives |
| Raw material manufacturer | |
| Product manufacturer | |

Related literature
1) G. O. Aspinall, E. L. Hirst, E. G. V. Percival, and I. R. Williamson: *J. Chem. Soc.* **1953**: 3184 (1953).
2) R. L. Whistler and E. L. Richards: *The Carbohydrates, Chemistry and Biochemistry*, 2nd. ed., vols. IIA, IIB, New York: Academic (1970), p. 447.

Chemical formula

| | |
|---|---|
| Name | Tamarind seed gum |
| Classification• network | Natural, organic, polysaccharides |
| Classification• crosslink point | Physical gel |
| Classification• fluid | Hydrogel |
| Manufacturing method | Extract the seed using hot water and by crushing it, from the albumen area of the bean-family vegetable tamarind, make a dry powder. |
| Solvent | Water (hot water, cold water) |
| Characteristics | A linear chain polymer has a molecular weight of about 115,000, and is a lemon colored powder with a grayish tint. No taste or odor. Disperses well in cold water and when heat is added becomes a consistent liquid. The solution shows properties of Newtonian fluid and is rather stable to heat. Gels under coexisting sugar and alcohol. |
| Gel preparation method | Gels when the concentration is above 0.5%, sugar content is 45–60, or it is placed in a sugar/alcohol mixture. Obtained by cooling or mixing with alcohol and solution that includes sugar after heating. |
| Uses | Food thickening agent, stable, gelling agent, sauce, pickled products, jellies, ice cream, etc. |
| Raw material manufacturer | Dainippon Pharmaceuticals |
| Product manufacturer | Dainippon Pharmaceuticals |
| Related literature | 1) M. Glicksman (ed.).: in *Food Hydrocolloids*, vol. III, Boca Raton, FL: CRC Press, (1986), p. 191. |
| | 2) P. Lang, and K. Kajiwara: *J. Biomater. Sci. Polym. Ed.* **4**: 517 (1993). |

Chemical formula

| Name | HS protein |
|---|---|
| Classification• network | Natural, organic, protein |
| Classification• crosslink point | Chemical gel |
| Classification• fluid | Hydrogel |
| Manufacturing method | After the wool is reduced, it is reacted with iodo acetic acid. When extracted using ammonium water, S-carboxymethyl keratin (SCMK) can be extracted. When fractionated according to solubility differences, it is divided into SCMKA with low solubility and SCMKB with high solubility. When SCMKB is fractionated further, HS protein is obtained. |
| Solvent | Ammonia water |
| Characteristics | The matrix consists of globular protein, with a molecular weight of around 20 kDa, which is rich in Cys (around 1100 µmol/g). |
| Gel preparation method | Reduce the hair using tri-n-butyl phosphine (TBP), and block the SH group using N-ethylmaleimide and swell it like a gel using 8-M LiBr/BC phosphine diluted products. |
| Raw material manufacturer | Canadian Biochemical Co. |
| Product manufacturer | |
| Related literature | 1) *Chemical Dictionary*, Kyoritsu Publ. |
| | 2) Japan Biochemistry Association (ed.): *Biochemistry Databook*, Tokyo Kagaku Dojin. |
| | 3) Japan Analytical Chemistry Association (ed.): *Polymer Analysis Handbook*, Kinokuniya Shobo. |
| Chemical formula | |

|  | Hard keratin (hair, nail, horn etc.) | | | Soft keratin (skin keratin layer) | |
|---|---|---|---|---|---|
|  | LS protein | HS protein | Glycin-rich protein | LSA protein | HS protein |
| Composition (%) | 40–85 | 5–45 | 1–30 | $\sim 93$ | $\sim 17$ |
| Molecular weight | $\sim 50{,}000$ | 10,000–30,000 | 5,000–10,000 | $\sim 60{,}000$ | 20,000–40,000 |
| Cysteine residue (%) | 4 | 10–30 | 5–11 | 1 | 2 |
| Glycine residue (%) | 3 | 3–9 | 20–40 | 14 | 17 |
| Helical content (%) | $\sim 50$ | 0 | 0 | $\sim 45$ | – |
| Origin | microfibril | matrix | matrix | fiber | unknown |

| Name | LS protein |
|---|---|
| Classification• network | Natural, organic, protein |
| Classification• crosslink point | |
| Classification• fluid | Hydrogel |
| Manufacturing method | Using wool, reduce using 0.3 MTGA solution (pH 11), which includes 8.0 M urea, neutralize it with acetic acid to pH 7, and oxidize it by adding 1.5 M NaBrO$_3$ aqueous solution under conditions in which there is sufficient unreacted TGA. It is obtained by filtering the undissolved material and dialating. Called CMADK. |
| Solvent | 0.3 M TGA water-soluble solution (pH 11) that includes 8.0 M urea |

| Characteristics | It has the characteristic of medium diameter filament (IF protein) of a eucaryotic cell. |
| --- | --- |
| | It can be divided into I-type with high acidic amino acids and neutral and basic II-type. The amount of Cys is small. |
| Gel preparation method | An opaque, coagulated gel is formed when a 4% CMADK aqueous solution is heated to 100°C. Swells in 8 M urea. |
| Uses | Reagent for immunological research |
| Raw material manufacturer | Canadian Biochemical Co. |
| Product manufacturer | |
| Related literature | 1) *Chemical Dictionary*, Kyoritsu Publ. |
| | 2) Japan Biochemistry Association (ed.): *Biochemistry Databook*, Tokyo Kagaku Dojin. |
| | 3) Japan Analytical Chemistry Association (ed.): *Polymer Analysis Handbook*, Kinokuniya Shobo. |
| Chemical formula | |

|  | Hard keratin (hair, nail, horn etc.) | | | Soft keratin (skin keratin layer) | |
| --- | --- | --- | --- | --- | --- |
|  | LS protein | HS protein | Glycin-rich protein | LSA protein | HS protein |
| Composition (%) | 40–85 | 5–45 | 1–30 | ~93 | ~17 |
| Molecular weight | ~50,000 | 10,000–30,000 | 5,000–10,000 | ~60,000 | 20,000–40,000 |
| Cysteine residue (%) | 4 | 10–30 | 5–11 | 1 | 2 |
| Glycine residue (%) | 3 | 3–9 | 20–40 | 14 | 17 |
| Helical content (%) | ~50 | 0 | 0 | ~45 | – |
| Origin | microfibril | matrix | matrix | fiber | unknown |

| Name | Tunicin |
| --- | --- |
| Classification• network | Natural, organic, polysaccharides |
| Classification• crosslink point | Physical gel |
| Classification• fluid | Hydrogel |
| Manufacturing method | Make a 3 mm square from the outer skin of Halocynthia roretzi, place this in 1 wt% NaOH solution at 96°C, stir for 6 h, and place it in ethanol at 70°C and refine it for 3 h. |
| Solvent | Water |
| Characteristics | The degree of crystallinity is high. |
| | Young's modulus = $6.64 \times 10^{10}$ dyn cm$^{-2}$, internal loss = 0.054 (density = 0.628 gcm$^{-3}$). |
| Gel preparation method | Unknown |
| Uses | Acoustic speaker vibration plate |
| Raw material manufacturer | Onkyo, Inc |
| Product manufacturer | |
| Related literature | 1) T. Nonaka and F. Horii: *Cellulose Communications* **3**: 41 (1996). |
| Chemical formula | Not listed |

| Name | Dextran |
|---|---|
| Classification• network | Natural, organic, polysaccharides |
| Classification• crosslink point | Chemical gel |
| Classification• fluid | Hydrogel |
| Manufacturing method | When a *Streptococcus* is cultured on a substrate that includes sucrose, a dextran made only from D-glucose is formed. Dissolve 25 g of external polysaccharide obtained from culture solution in one liter of water, and add ethanol until it reaches 45% concentration; then precipitate the dextran. The precipitated material is again dissolved in water and is refined by repeatedly precipitating it in an equal volume of ethanol. |
| Solvent | Water |
| Characteristics | The gel-like material Sephades made by crosslinking dextran with epichlorohydrin is used as molecular sieve for materials with molecular weight of 700–200,000 Daltons, depending on the degree of crosslinking. It forms glucose or isomaltose oligosaccharide by enzyme (dextranase). No color reaction due to iodine. $[\alpha]_D = +215°$ |
| Gel preparation method | Dextran is crosslinked with epichlorohydrin |
| Uses | Dextran sulfate is used as a blood coagulation prevention agent. When the average molecular weight is adjusted to $75,000 \pm 25,000$ Daltons, it serves as a blood substitute to increase blood level. |
| Raw material manufacturer | |
| Product manufacturer | Nakarai Tesque |
| Related literature | 1) A. F. Charles and L. N. Farrell: *Can. J. Microbiol.* **3**: 329 (1957).<br>2) A. F. T. Gronwall and B. G. A. Ingelman: *Acta Physiol. Scand.* **7**: 97 (1944). |
| Chemical formula | |

| Name | Dermatan sulfate |
|---|---|
| Classification• network | Natural, organic, polysaccharides |
| Classification• crosslink point | |
| Classification• fluid | |
| Manufacturing method | Alkali extracted from pig skin, and after fractionation by protease and alcohol, refine using Schiller's chromatography. |
| Solvent | Water |
| Characteristics | $[\alpha]_D^{20} = -67--75°$, molecular weight: 11,000–25,000, stops blood coagulation. |
| Gel preparation method | Unknown |
| Uses | |
| Raw material manufacturer | Biochemical Co. |
| Product manufacturer | Biochemical Co. |
| Related literature | 1) K. Abu and N. Seni: *Basics of Sugar Chemistry*, Kodansha (1984), p. 147. |

Chemical formula

| Name | Starch |
|---|---|
| Classification• network | Natural, organic, polysaccharides |
| Classification• crosslink point | Physical gel |
| Classification• fluid | Hydrogel |
| Manufacturing method | After peeling potatoes, soak in 0.1% pyrosulfite solution. Cut into pieces of about 1 cm and crush in a blender for 1 min. Filter through a cloth with ~0.2 mm grain. After leaving the filtrate for several hours, starch will precipitate to the bottom. |
| | Soluble starch is obtained by suspending starch in 0.5–10% mineral acids. It is obtained by soaking starch in 7.5% hydrochloric acid at room temperature for 7 days, and then washing and drying it. |
| Solvent | Water |
| Characteristics | Made from amylose and amylopectin, and the percentages of both differ depending on the type of starch. |
| Gel preparation method | When dissolving in water, it becomes a sol with a blue tint, and when the concentration is high and it cools, it hardens into a white gel. |
| Uses | As a derivative, there is dextrin and acidic starch, which is used for the paper and textile-making industries. |
| | Food, cosmetics, clothing, stamps. |
| Raw material manufacturer | E. Merck Darmstadt |
| Product manufacturer | Nakarai Tesque |
| Related literature | 1) J. N. Bemiller: *Starch Amylose in Industrial Gums*, 2nd ed., R. L. Whistler (ed), New York: Academic Press (1973), pp. 545–566; E. L. Powell: *Starch Amylopection*, New York: Academic, pp. 567–576. |
| | 2) M. H. A. De Willigen: *Methods in Carbohydrate Chemistry*, R. E. Whistler and J. N. Be Miller (eds), vol. IV, New York: Academic Press (1964), p. 9. |

Chemical formula

| Name | Tragacanth gum |
|---|---|
| Classification• network | Natural, organic, polysaccharides |
| Classification• crosslink point | |
| Classification• fluid | Hydrogel |
| Manufacturing method | Scar the bark of the *Astragalus* tree and collect the rubber that seeps out from that opening, then dry it. This is manufactured into ribbon shapes or thin layers (*Astragalus* grows near the dry and mountainous regions of the Black Sea, Saudi Arabia, Iran, Syria, and Turkey). |

| | |
|---|---|
| Solvent | Water |
| Characteristics | The molecular weight of the linear polymer is 840,000. It is a long structure, molecule size of 450 nm, 0.9 nm width. Tragacanth gum has at least two types of polysaccharides: the insoluble bassorin and the water-soluble tragacanthin. Normally the gum consists of 70% tragacanthin (swollen area) and 10% araban (dissolvable area), 10% water, 4% cellulose, 3% starch, and 3% ash. When tragacanth gum is dissolved in water, the soluble tragacanthin becomes a colloidal hydrophilic sol, but the nonsoluble bassorin swells and becomes a gel. |
| Gel preparation method | When dissolved in water, it becomes a lump. It is difficult to obtain a homogeneous solution, so it is added slowly while being quickly stirred or it is wetted with glycerin, propylene alcohol, or alcohol before dissolving. When freeze-dried material is dissolved, it swells more easily and dispersion improves. |
| Uses | Manufactured as emulsifiers. Used in cosmetics and drinks. |
| Raw material manufacturer | |
| Product manufacturer | Kato Yoko, Nagase Industries (import handling agent) |
| Related literature | 1) F. Smith: *J. Chem. Soc.* 274 (1939). |
| | 2) J. Jackson and F. Smith: *J. Chem. Soc.* 74 (1940). |
| | 3) S. P. James and F. Smith: *J. Chem. Soc.* 739 (1945). |
| | 4) S. W. Challinor, W. N. Haworth, and E. L. Hirst: *J. Chem. Soc.* 258 (1931). |
| Chemical formula | |

---

| | |
|---|---|
| Name | Nylon |
| Classification• network | Synthetic, organic, polysaccharides |
| Classification• crosslink point | Chemical gel |
| Classification• fluid | Organogel |
| Manufacturing method | 6-nylon: ring-opening polymerization of ε-caprolactam |
| | 6,6-nylon: condensation polymerization of hexamethylene diamine and adipic acid |
| | 6,10-nylon: condensation polymerization of hexamethylene diamine and sebacic acid. |
| Solvent | Phenol, m-cresol, formic acid |
| Characteristics | A semiclear solid, with impact, heat and wear resistance and has little friction. Does not dissolve in normal organic solvents. Has a strong resistance to alkali, but dissolves in strong acids and hydrolyzes. Becomes brittle under ultraviolet light. Dissolves at normal temperature in phenol, and it dissolves ethylene glycol and benzyl alcohol when heated. There is relatively low solvent affinity and the cross-linking density is high. Clothing is made using fibers obtained by melt-spinning. Various types of nylons. |
| Gel preparation method | There is an example of crosslinking and gelling the surface in contact with a plasma from an inert gas using high-frequency electical discharge |
| Uses | Clothing, tire cords, belt cords, brushes, gears, bearings, artificial skin, covering for electrical lines. |

| | |
|---|---|
| Raw material manufacturer | Toray Industries, Inc., Toyo Boseki, Asahi Glass, Yunichika, Kanebo, Teijin |
| Product manufacturer | Related documents manufactured and sold by various chemical and textile manufacturers. |
| Related literature | |
| Chemical formula | |

$$+CO(CH_2)_6NH+_n \qquad \text{6-nylon}$$
$$+NH(CH_2)_6NHCO(CH_2)_4CO+_n \qquad \text{6,6-nylon}$$
$$+NH(CH_2)_6NHCO(CH_2)_8CO+_n \qquad \text{6,10-nylon}$$

---

| | |
|---|---|
| Name | Nigeran |
| Classification• network | Natural, organic, polysaccharides |
| Classification• crosslink point | Physical gel |
| Classification• fluid | Hydrogel |
| Manufacturing method | Extracted from suspended liquid obtained by fracturing after culturing *Aspergillus niger* in medium that includes maltose. |
| Solvent | Water |
| Characteristics | Dissolves in hot water, precipitates in cold water. |
| | $[\alpha]_D = +283°$ |
| | $M_w = \sim 5 \times 10^4$ |
| Gel preparation method | Gels after being dissolved in hot water and then cooled. |
| Uses | |
| Raw material manufacturer | SIGMA |
| Product manufacturer | |
| Related literature | 1) K. K. Tumg and J. H. Nordin: *Biochim. Biophys. Acta.* **158**: 154 (1968). |
| | 2) S. A. Braker and T. R. Carrington: *J. Chem. Soc.* 3588 (1953). |
| Chemical formula | |

---

| | |
|---|---|
| Name | Hyaluronic acid [HA] |
| Classification• network | Natural, organic, polysaccharides |
| Classification• crosslink point | Chemical gel |
| Classification• fluid | Hydrogel |
| Manufacturing method | Hyaluronic acid is isolation-refined from cockscomb and animal organs. It is also taken from microorganisms, and two bacteria, *Streptococcus equi* and *Streptococcus zooepidemicas* from the Lancefield group C are often used. When the bacteria grows, hyaluronic acid and lactic acid are formed at the same time, but when the pH of the medium decreases, hyaluronic acid formation is inhibited, so the pH of the medium should be neutral. Normally, the incubation is completed in 30–40 h and after that it is disinfected using heat or chemicals and then the nonsoluble materials or the bacteria are removed through filtration or centrifugation. Methods using activated carbon treatment and alcohol precipitates yield a high-purity hyaluronic acid. |

| | |
|---|---|
| Solvent | |
| Characteristics | Hyaluronic acid is a linear polymer bonded alternatively by N-acetyl-D-glucosamine residues and D-glucosamine acid residues. It is one of the glucosaminoglycans, with molecular weight of from several 10,000 to several millions. It is seen in joint fluids in animals, the vitreous of the eye, ligaments and corium layer. The molecular weight of the sugar chain is $10^5$–$10^6$, and it is rather long, but differs depending on the material. The length is not the same even among the same materials. It forms a gel after bonding with a large amount of water, and can lubricate joints inside an organism, giving skin flexibility. |
| Gel preparation method | Mix with sodium hyaluronic acid with NaOH to produce hyaluronic acid solution. It is deaerated to make it a homogeneous solution (excluding the foam). Add the crosslinking agent with cold water, mix well, and let it react for 5 to 6 h at 35–40°C. |
| Uses | Used as raw materials for moisturizer of cosmetics and medicines (used in eye-surgery, knee-joint lubricant, wrapped-medicinal capsules), used for medical tools. |
| Raw material manufacturer | Shiseido |
| Product manufacturer | Wako Pure Chemical Industries |
| Related literature | 1) K. Takemoto, J. Sunamoto, and M. Akashi (eds.): *Polymers and Medicine*, Mitsuda Publ. (1989), p. 356. |
| | 2) K. Hagino, Y. Osada, T. Fushimi, and A. Yamauchi (eds.): Sangyo Tosho, *Gels-Basics and Applications of Soft Material*, p. 26, 165. |
| Chemical formula | |

---

| | |
|---|---|
| Name | Bis (acrylamidemethyl) ether |
| Classification• network | Synthetic, organic, vinyl polymer |
| Classification• crosslink point | Chemical gel |
| Classification• fluid | Hydrogel |
| Manufacturing method | |
| Solvent | Water |
| Characteristics | The gel copolymerized with acrylamide and N-methacrylamide is used in electrophoresis. The automatic DNA sequences-decoding ability of this gel spans 1000 basic groups. |
| Gel preparation method | Unknown |
| Uses | Crosslinking agent (for electrophoresis) |
| Raw material manufacturer | AT Biochem |
| Product manufacturer | Hoshuzo |
| Related literature | 1) R. J. Molinari *et al.*: *Adv. Electrophoresis* **6**: 43–60 (1993). |

Chemical formula

$$CH_2 = CH$$
$$|$$
$$C=O$$
$$|$$
$$N-H$$
$$|$$
$$CH_2$$
$$|$$
$$O$$
$$|$$
$$CH_2$$
$$|$$
$$N-H$$
$$|$$
$$C=O$$
$$|$$
$$CH_2 = CH$$

| | |
|---|---|
| Name | Hydroxyethyl cellulose [HEC] |
| Classification• network | Natural, organic, polysaccharides |
| Classification• crosslink point | |
| Classification• fluid | Hydrogel |
| Manufacturing method | Obtained by addition of ethylene oxide or ethylene chlorohydrin to alkali cellulose under pressure. There are times when the alkylene oxide is added to the hydroxylalkyl group that is added to the cellulose. This means not all bonded alkylene oxide is always directly bonded to the cellulose molecule. |
| Solvent | Water (see characteristics) |
| Characteristics | Linear polymer is a white powder with no taste or odor and has a specific gravity of 0.55–0.75. The aqueous solution has a pH of 6.5–8.5. When the attached mols of ethylene oxide or ethylene chlorohydrin are >0.5, it dissolve in warm or cold water. This solution is little affected by inorganic salts and has a wide range of application pH. |
| Gel preparation method | Crosslinked under acid using cellulose crosslinking agents such as dialdehyde and aldehyde (formaldehyde). It polymerizes and crosslinks with a vinyl monomer and crosslinking agent (methylene bis acrylamide) using the free radical initiator. |
| Uses | Has exceptional emulsion dispersion, stabilization, and viscosity thickening effects, and is used to process and control the viscosity of shampoos and lotions. It moisturizes and controls hair so it is added to setting lotions and hair dyes. It is also used as a bonding agent in toothpaste and makeup. |
| Raw material manufacturer | Fuji Chemical, importer = Nissho Iwai [Manufacturer: Hercules Co. (U.S.A.)], Daicel Chemical [Manufacturer: UCC (U.S.A.)] |
| Product manufacturer | |
| Special note | Manufacturing has currently stopped, and it is difficult to obtain. |
| Related literature | |
| Chemical formula | |

$$\left[ \begin{array}{c} CH_2OX \\ H \quad O \\ H \\ OY \quad H \\ H \quad H \\ H \quad OZ \end{array} -O- \right]_n$$

X : ( $CH_2CH_2O$ )$_x$ − H or –H
Y : ( $CH_2CH_2O$ )$_y$ − H or –H
Z : ( $CH_2CH_2O$ )$_z$ − H or –H

| | |
|---|---|
| Name | 12-Hydroxystearic acid |
| Classification• network | Synthetic, organic, low molecular weight material. |
| Classification• crosslink point | Physical gel |
| Classification• fluid | Organogel |
| Manufacturing method | Either process the oleic acid with concentrated sulfuric acid, or obtain by reducing with 10-sodium oxostearic acid. |
| Solvent | Vegetable oil, cyclohexane, chloroform |
| Characteristics | MP (melting point) 82°C |
| Gel preparation method | Dissolve in solvent where the temperature is above the melting point (around 80°C) and cool to under 40°C. By mixing with polar material such as lecithin, a decrease in gel strength occurs and gels may not form. |
| Uses | Domestic waste-oil treatment agent |
| Raw material manufacturer | Wako Pure Chemical Industries |
| Product manufacturer | Wako Pure Chemical Industries |
| Special note | CAS Registry Number [106-14-91] |
| Related literature | 1) T. Tamura, *Hyomen*, **32**, 73 (1995). |

| | |
|---|---|
| Name | Hydroxypropyl cellulose [HPC] |
| Classification• network | Natural, organic, polysaccharides (modified process) |
| Classification• crosslink point | Physical gel |
| Classification• fluid | Organogel, hydrogel |
| Manufacturing method | Made by reacting propylene oxide with alkali cellulose. |
| Solvent | Polar organic solvent, water |
| Characteristics | Linear polymer softens at 130°C. Precipitated from water at 40–45°C |
| Gel preparation method | Crosslinks with cellulose crosslinking agents (formaldehyde, dialdehyde, diisocyanate, halohydrin). It will also graft polymerize and crosslink vinyl polymers and crosslinking agents (MBAA, methylene bis acrylamide) by free radical initiators. |
| Uses | Emulsion, cosmetics, coating for pill-releasing agents |
| Raw material manufacturer | |
| Product manufacturer | Wako Pure Chemical Industries |
| Related literature | 1) K. D. Klug *J. Polym. Sci.* C, **36**: 491 (1971). |
| Chemical formula | |

$R = CH_2CH(OH)CH_3$

| | |
|---|---|
| Name | Fucoidan |
| Classification• network | Natural, organic, polysaccharides |
| Classification• crosslink point | |
| Classification• fluid | |
| Manufacturing method | Extracted using hot acid at pH = 2 or hot water from *Fucus vesiculosus* and bonded with quaternary ammonium salts such as setaprone or CPC, and then separated. |

Solvent | Hot water
Characteristics | $[\alpha]_D = -75\text{--}140°$, $M_w = 7.8 \times 10^4\text{--}1.53 \times 10^5$
Anticoagulant for blood, clearing agent for platelets, and high hygroscopicity.
Gel preparation method | Unknown
Uses |
Raw material manufacturer | SIGMA
Product manufacturer |
Related literature | 1) W. A. Black: *Chem. Process Chicago* **16**: 139 (1953).
2) A. N. O'Neill: *J. Am. Chem. Soc.* **76**: 5074 (1954).

Chemical formula

---

Name | Pustulan
Classification• network | Natural, organic, polysaccharides
Classification• crosslink point | Physical gel
Classification• fluid | Hydrogel
Manufacturing method | After extracting from lichen, *Umbilicaria pustulata*, the acetyl group that is included in around 10% of the glucose residues is eliminated in a dilute sodium carbonate aqueous solution and is precipitation-purified using isopropanol.
Solvent | Water
Characteristics | $M_w = 2 \times 10^4\text{--}5 \times 10^4$, $[\alpha]_D = -10\text{--}-33°$.
When heated at 80°C, it transfers from a gel to liquid.
Gel preparation method | The pustulan that was deacetylated due to the sodium carbonate aqueous solution is made into a solution of 0.5–10 wt%, and heated to 95–100°C. After cooling to room temperature, it is left for several hours to several days.
Uses |
Raw material manufacturer | Calbiochem-Behring
Product manufacturer |
Related literature | 1) A. J. Stipanovic and P. J. Giammatteo: *Industrial Polysaccharides*: *Genetic Engineering, Structural Property Relations and Applications*, M. Yalpani (ed), New York: Elsevier Science Publishers (1987), p. 281.
2) A. J. Stipanovic, P. J. Giammatteo, and S. B. Robie: *Biopolymers* **24**: 2333 (1985).

Chemical formula

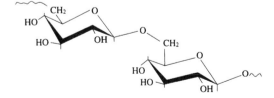

| Name | Funoran |
|---|---|
| Classification• network | Natural, organic, polysaccharides |
| Classification• crosslink point | Physical gel |
| Classification• fluid | Hydrogel |
| Manufacturing method | Extract from *Gloiopeltis furcata* under acidic conditions, using hot water. Fractionate the precipitate formed by adding the sodium cetyl pyridium using KCl solution. |
| Solvent | Water |
| Characteristics | Has a structure close to agarose. |
| | L-Gal residue is not included and has a significant sulfate group 16% |
| Gel preparation method | Add organic solvents such as DMSO to solution |
| | Makes a weak gel when eluting a cetyl pyridium complex with 2% $KCl_{aq}$ |
| Uses | Stabilizer or dispersing agent for medicine |
| | Raw materials for chewing gum, pills, drinks, ice cream, chocolate |
| | Used also for toothpaste, mouthwash, and oral refreshers. |
| Raw material manufacturer | |
| Product manufacturer | |
| Related literature | 1) S. Hirase, C. Araki, and T. Ito: *Bull. Chem. Soc. Japan* **31**: 428 (1958). |
| | 2) M. Watase and K. Nishinari: *Polym. J.* **20**: 1125 (1988). |
| | 3) *Polysaccharides Science*, vol. 2, Kodansha. |
| Chemical formula | |

| Name | Pullulan |
|---|---|
| Classification• network | Natural, organic, polysaccharides |
| Classification• crosslink point | Physical gel |
| Classification• fluid | Hydrogel |
| Manufacturing method | Cultured by adding glucose or cane sugar in a Czapek-Dox medium |
| | Adding vitamin B1 allows higher yield. |
| Solvent | Water |
| Characteristics | $[\alpha]_D = +190\text{--}192°$, $M_w = 1.5\text{--}4 \times 10^6$ |
| | Hydrolyze to maltotriose by pullulanase |
| | Has high membrane-creating ability, adhesive ability, viscoelasticity, and high processability |
| Gel preparation method | When dissolved in cold or warm water, it becomes a solution with strong adhesion. |
| | It does not have an ability to gel, but when the solution is dried, a film is formed. |
| Uses | Film materials and plastic materials |
| | Sheet foods, coating materials, binding agents |
| Raw material manufacturer | Hayashihara Bio-organism Chemistry Research Center, Biochemical Co. |
| Product manufacturer | |
| Special note | Store at 4°C. $M_n = 5 \times 10^4\text{--}1 \times 10^5$ |

Related literature

1) K. Wallenfels, H. Bender, G. Keilich, and G. Bechtler: *Angew. Chem.* **73**: 245 (1951).
2) K. Wallenfels, G. Keilich, G. Bechtler, and D. Freudenberger: *Biochem. Z.* **341**: 433 (1965).

Chemical formula

| Name | Degraded xyloglucan |
|---|---|
| Classification• network | Natural, organic, polysaccharides |
| Classification• crosslink point | Physical gel |
| Classification• fluid | Hydrogel |
| Manufacturing method | The $\beta$-galactosidase is used to partially decompose the side chain galactose of xyloglucan polysaccharide. |
| Solvent | Water (cold water) |
| Characteristics | Xyloglucan from tamarind gels when over 30% of the galactose is removed. It gels when heated and when cooled, and it reversibly sols. The sol-gel transition temperature and the physical properties differ according to the galactose removal rate. |
| | This gel has some elasticity and is smooth. |
| | It is considered that gelation takes place due to the association of molecular chains caused by a hydrophobic bond. |
| Gel preparation method | It gels when heat is added after dissolving under freezing temperature and agitating. |
| Uses | Food gelling agent, stabilizer. Used in jellies and microwavable foods. Chemical release control agents. |
| Raw material manufacturer | Dainippon Pharmaceuticals |
| Product manufacturer | Dainippon Pharmaceuticals |
| Related literature | 1) M. Shirokawa and K. Yamatotani: *Japan Agricultural Chemistry Association Documents*, p. 182 (1995). |
| | 2) Y. Yuguchi *et al.*: *Green Polymers*, JETRO (1997). |

Chemical formula

| Name | High methoxy pectin |
|---|---|
| Classification• network | Natural, organic, polysaccharides |
| Classification• crosslink point | Physical gel |
| Classification• fluid | Hydrogel |
| Manufacturing method | Extracted from apples and citrus fruits (lemons, limes). |
| Solvent | Water |
| Characteristics | Forms gels that are elastic and have low pH under the existence of sugars. |
| | Forms gels that have a strong acid resistance. |
| | Due to changes in sugar level and esterification the gelation rate differs (the higher the esterification and the sugar rate, the higher the gelling temperature and speed). The crosslinking structure of the gel is formed by the association of ribbon structures. |
| Gel preparation method | Forms when sugar content is over 55%, low pH (2.5–4.0). Gel is made when solution is cooled after it has been heated. |
| Uses | Food gels, stabilizers, jams, jellies, yogurt |
| Raw material manufacturer | SBI, Copenhagen Pectin, H&F, Citrus Colloid |
| Product manufacturer | Yukijirushi Foods, Sansho, Dainippon Pharmaceuticals, Morishita Industries |
| Related literature | 1) E. R. Morris *et al.*: *Int. J. Biol. Macromol.* **2**: 327 (1980). |

Chemical formula

| | |
|---|---|
| Name | Low methoxy pectin |
| Classification• network | Natural, organic, polysaccharides |
| Classification• crosslink point | Physical gel |
| Classification• fluid | Hydrogel |
| Manufacturing method | Obtained by demethylating using acid or alkali after extracting from apples and citrus fruit (lemons, limes). |
| Solvent | Water |
| Characteristics | Forms an elastic gel under the existence of a bivalent cation. Forms a strong, acid-resistant gel. Coordination bonded by a divalent cation to the area where it is not esterified in the pectin molecular chain. |
| Gel preparation method | Gels when solution is heated then cooled under the existence of a divalent cation. The amount of cation changes based on pH and the degree of esterification. Gel can be obtained even when a pectin solution and divalent ion solution is mixed at low temperatures. |
| Uses | Food gels, stabilizers, jams, milk-desserts, jellies |
| Raw material manufacturer | SBI, Copenhagen H&F, Obi |
| Product manufacturer | Yukijirushi Foods, Sansho, Dainippon Pharmaceuticals, Tomen |
| Related literature | 1) E. R. Morris *et al.*: *Int. J. Biol. Macromol.* **2**: 327 (1980). |
| Chemical formula | |

| | |
|---|---|
| Name | Poly(acrylamide) [AAm], [PAAm] |
| Classification• network | Synthetic, organic, vinyl polymer |
| Classification• crosslink point | Chemical gel |
| Classification• fluid | Hydrogel |
| Manufacturing method | Acrylamide is polymerized by ultraviolet light, heat, or peroxide in water or an organic solvent. |
| Solvent | Easily dissolves in water, alcohols, esters, and ketones |
| Characteristics | It is a scale-like crystal with no color and a melting point of 84.5°C. It shows high solubility to polar solvents other than water-like alcohol. If it is pure, then it can be stored for several months in a dark place without stabilizing agents. The polymer's concentrated aqueous solutions show gel-like properties. Partially hydrolyzed gel will exhibit volumetric phase transition by changing the temperature, pH, and solvent composition. |

| | |
|---|---|
| Gel preparation method | It is possible to solid-state polymerize by radical polymerization, anionic polymerization, and $\gamma$-ray irradiation. As a crosslinking agent, it is possible to gel using multifunctional groups (N,N-methylene-bis-acrylamide, diepoxy). It is also possible to obtain various types of gels including ones that are ionized and gels that have been neutralized and do not contain ions, by including comonomers and by hydrolysis. |
| Uses | Aggregating agents, soil improvement agents, polymer processing or improving textiles, paper strengthening agent, adhesive, paints, oil salvaging agents. |
| Raw material manufacturer | Mitsubishi Chemical, Mitsui Toatsu Chemical |
| Product manufacturer | |
| Special note | This is a poisonous material as defined by Labor and Sanitation Laws, so it is necessary to use caution when handling |
| Related literature | 1) Organic Synthetic Chemistry Association: *Organic Compounds Dictionary*, Kodansha (1985). |
| | 2) M. Irie (ed.): *Manufacturing and Application of Functional Polymer Gels*, CMC, p. 28 (1987). |
| | 3) Chemistry Dictionary Editorial Association (ed.): *Chemistry Dictionary*, Kyoritsu Publ. (1963). |
| Chemical formula | |

$$\mathrm{+CH_2-CH\,+_n}$$
$$|$$
$$\mathrm{CONH_2}$$

---

| | |
|---|---|
| Name | Poly(2-acrylamido-2-methylpropane sulfonic acid) [PAMPS] |
| Classification• network | Synthetic, organic, vinyl polymer |
| Classification• crosslink point | Chemical gel |
| Classification• fluid | Hydrogel |
| Manufacturing method | React acrylonitrile, isobutane and fuming sulfuric acid. A polymer is obtained by free radical polymerization of 2-acrylamide-2-methylpropyl sulfonic acid in water. |
| Solvent | Water, DMF |
| Characteristics | Molecular weight of monomer is 207.246. White crystal with the taste and odor of an acid. Melting point is 185°C (there is a possibility of partial decomposition). |
| | Shows a strong acidic tendency and the pH depends on the concentration. |
| Gel preparation method | Radical polymerization under water, under the existence of cross-linking agents. |
| | Polymerization is possible in nonaqueous systems if it is made into an amine salt. |
| Uses | Dye enhancers, aggregators, worm repellents, paint, photography film, cosmetics, contact lenses |
| Raw material manufacturer | Nitto Chemical, Lubrizol, Toa Gosei Corp |
| Product manufacturer | |
| Special note | It does cause a slight hydrolysis, but this can be prevented by sodium salt. Monomer has sudden oral toxicity. |
| | LD50 $= 5$–$10\,\mathrm{g/kg}$ |
| Related literature | |

Chemical formula

$$\{CH_2-CH\}_n$$
$$|$$
$$C-NH-CH(CH_3)-CH_2-SO_3H$$
$$\parallel \qquad |$$
$$O \qquad CH_3$$

| | |
|---|---|
| Name | Poly(acrylic acid) |
| Classification• network | Synthetic, organic, vinyl polymer |
| Classification• crosslink point | Chemical gel |
| Classification• fluid | Hydrogel |
| Manufacturing method | Polymer is obtained by free radical polymerization of acrylic acid. Direct oxidation method of propylene |
| Solvent | Water, hydrophilic alcohol-water, acetone |
| Characteristics | Monomer's molecular weight; 72.03. Liquid that looks like clear acetic acid (coagulation point 12°C). Specific gravity, 1.1, boiling point 141.7°C. Normally, hydroquinone is added as a polymerization inhibitor, but at high temperature, it is easy to polymerize. The degree of swelling and solubility for those that have polymerized or gelled will change in the presence of acid and salt. |
| Gel preparation method | Mix the appropriate amount of solvent, acrylic acid, polymerization initiator, and crosslinking agent and after removing the oxygen in the solution using vacuum deaeration, let it sit for $\sim$12 h at 60°C. After polymerization, it will reach equilibrium swelling by immersing the solvent for around a week. |
| Uses | Thickeners, pap agents |
| Raw material manufacturer | Toa Gosei Corp |
| Product manufacturer | Japan Catalyst |
| Related literature | |
| Chemical formula | |

$$\{CH_2-CH\}_n$$
$$|$$
$$COOH$$

| | |
|---|---|
| Name | Poly(N,N-dimethylaminoethyl acrylate) |
| Classification• network | Synthetic, organic, vinyl polymer |
| Classification• crosslink point | Chemical gel |
| Classification• fluid | Hydrogel, organogel |
| Manufacturing method | Polymer is obtained by free radically polymerizing dimethyl ethyl ester acrylic acid in water or an organic solvent. For polymerizing in solution, make sure it has been made into a quaternized salt. |
| Solvent | Water, alcohol, acetic acid ethyl, benzene, xylene, carbon tetrachloride |
| Characteristics | Molecular weight of the monomer is 143.2, boiling point is 75°C/mmHg, melting point is 75°C, specific gravity is 0.943 (20/4°C), refractive index is 1.4373 ($n$ D$^{20}$), and flashpoint is 63°C. Outer appearance has no color, or a brownish clear liquid. Viscosity 1.34 cP. |

| | |
|---|---|
| Gel preparation method | Gel is obtained by radical polymerization under the existence of crosslinking agents. |
| Uses | Homopolymer and copolymer are used to improve dyeing and adhesive abilities. |
| Raw material manufacturer | Kyojin, Inc. |
| Product manufacturer | |
| Special note | A highly pure polymer is obtained from polymerization of an aqueous solution of methacrylic acid ester. Monomer has oral toxicity. LD50 = 2300 mg/kg (mouse) |
| Related literature | 1) Japanese Unexamined Patent Application publication 3-81310. |
| | 2) Japanese Unexamined Patent Application publication 58-1707 |
| Chemical formula | |

$$\begin{array}{c} \text{\textonehalf}CH_2-CH\text{\textonehalf}_n \\ | \\ COO-(CH_2)_2-N-CH_3 \\ | \\ CH_3 \end{array}$$

| | |
|---|---|
| Name | Poly(sodium acrylate) |
| Classification• network | Synthetic, organic, vinyl polymer |
| Classification• crosslink point | Chemical gel, physical gel |
| Classification• fluid | Hydrogel |
| Manufacturing method | Aqueous radical polymerization of sodium acrylic acid. |
| Solvent | Water |
| Characteristics | It is a white powder. |
| | A clear and consistent solution is created when dissolved in water. |
| Gel preparation method | The concentrated, thick water-soluble solution itself is the gel, but it becomes a gel by crosslinking with polyvalent metallic salt. |
| | Add diepoxy compound, polyvalent amine, polyvalent alcohol, and alkylene carbonate, then heat and form the gel. |
| Uses | Aggregating agents, thickening agents, pap agents |
| | Food additives |
| Raw material manufacturer | Toa Gosei Corp |
| Product manufacturer | Japan Catalyst |
| Related literature | |
| Chemical formula | |

$$\begin{array}{c} \text{\textonehalf}CH_2CH\text{\textonehalf}_n \\ | \\ COONa \end{array}$$

| | |
|---|---|
| Name | Crosslinked sodium polyacrylate |
| Classification• network | Natural, organic, vinyl polymer |
| Classification• crosslink point | Chemical gel |
| Classification• fluid | Hydrogel |

Manufacturing method
It is obtained as a dry powder through arrangement gel by reverse-phase suspension polymerization or aqueous free radical polymerization when acrylic acid or sodium acrylic acid mixed solution is under the existence of a small amount of crosslinking monomer that is a multifunctional acrylate.

It can also be obtained by mixing a small amount of polyvalent metallic salt, polyepoxy compounds, polyfunctional amine and polyfunctional alcohol with a solution of polyacrylic acid. It is obtained by drying this aqueous gel.

Solvent
Water

Characteristics
It is a white powder.

It absorbs water and swells when it comes in contact with water.

Gel preparation method
A hydrogel is obtained by dispersing powder in water or a hydrophilic solvent. It is possible to obtain a hydrogel using the same processes as for the manufacturing method.

Uses
Sanitary products (disposable diapers, sanitary towels) to absorb body fluids, pet sheets (urine absorbent materials), agent to maintain freshness, cold insulator, agriculture and gardening moisturizing agent, filler for stopping water, medical waste-fluid coagulator.

Raw material manufacturer
Japan Catalyst, Mitsubishi Chemical, Toa Gosei Co., BASF, R&H, others

Product manufacturer
Japan Catalyst, Sanyo Chemical, Sumitomo Seika, Kao Corp. Mitsubishi Chemical, Stockhausen, Dow, Chemdol, BASF, NAII, others

Related literature

Chemical formula

$$-(CH_2CH)---- \quad -(CH_2CH)--- \quad -(CH_2CH)---- \quad -(CH_2CH)-$$
$$\quad\quad | \quad\quad\quad\quad\quad | \quad\quad\quad\quad\quad | \quad\quad\quad\quad\quad |$$
$$\quad\quad COOH \quad\quad\quad COONa \quad\quad\quad CO \quad\quad\quad\quad COONa$$
$$\quad\quad\quad\quad\quad\quad\quad\quad\quad\quad\quad\quad\quad | $$
$$\quad\quad\quad\quad\quad\quad\quad\quad\quad\quad\quad\quad\quad X$$
$$\quad\quad\quad\quad\quad\quad\quad\quad\quad\quad\quad\quad\quad |$$
$$\quad\quad\quad\quad\quad\quad\quad\quad\quad\quad\quad\quad\quad CO$$
$$\quad\quad\quad\quad\quad\quad\quad\quad\quad\quad\quad\quad\quad |$$
$$-(CH_2CH)---- \quad -(CH_2CH)---- \quad -(CH_2CH)---- \quad -(CH_2CH)-$$
$$\quad\quad | \quad\quad\quad\quad\quad | \quad\quad\quad\quad\quad\quad\quad\quad\quad |$$
$$\quad\quad COOH \quad\quad\quad COONa \quad\quad\quad\quad\quad\quad\quad\quad COONa$$

---

Name
Poly(N-acryloylaminoethoxyethanol)

Classification• network
Synthetic, organic, vinyl polymer

Classification• crosslink point
Chemical gel

Classification• fluid
Hydrogel

Manufacturing method
React acryloylchloride with aminoethoxyethanol under basic conditions (Schotten–Baumann reaction).

Solvent
Water

Characteristics
Water-soluble polymer.

Polymer gel is stable over 500 times for hydrolysis in comparison to polyacrylamide gel.

Shows a high resolution to DNA analysis.

Radical polymerization is done by N,N'-methylene bis acrylamide for crosslinking agent, ammonium persulfate as an initiator, N,N,N',N'-tetramethylethylenediamide (TEMED) as an accelerator.

Uses      Used in electrophoresis (2-chain DNA analysis)

Raw material manufacturer      Aldrich

Product manufacturer

Related literature

1) M. Chiari *et al.*: *Electrophoresis* **15**: 177–186 (1994).
2) M. Chiari *et al.*: *Electrophoresis* **15**: 616–622 (1994).
3) M. Chiari *et al.*: *Electrophoresis* **16**: 1815–1829 (1995).
4) E. Simo-Alfonso *et al.*: *Electrophoresis* **17**: 723–731 (1996).

Chemical formula

$$\left.\!\!\!-\!\!\left(\, CH_2 \!-\!\! CH \,\right)\!\!\!\right._{n}$$
$$\begin{array}{c} | \\ C=O \\ | \\ N-H \\ | \\ CH_2 \\ | \\ CH_2 \\ | \\ CH_2 \\ | \\ OH \end{array}$$

| | |
|---|---|
| Name | Poly(N-acryloylaminopropanol) |
| Classification• network | Synthetic, organic, vinyl polymer |
| Classification• crosslink point | Chemical gel |
| Classification• fluid | Hydrogel |
| Manufacturing method | React acryloylchloride with aminoethoxyethanol under basic conditions (Schotten–Baumann reaction). |
| Solvent | Water |
| Characteristics | Water-soluble polymer. |
| | Polymer gel is stable over 500 times for hydrolysis in comparison to polyacrylamide gel. |
| | Shows a high resolution to DNA analysis. |
| Gel preparation method | Radical polymerization is done by N,N'-methylene bis acrylamide for crosslinking agent, ammonium persulfate as an initiator, N,N,N',N'-tetramethylethylenediamide (TEMED) as an accelerator. |
| Uses | Used in electrophoresis (2-chain DNA analysis) |
| Raw material manufacturer | Aldrich |
| Product manufacturer | |
| Related literature | 1) M. Chiari *et al.*: *Electrophoresis* **16**: 1815–1829 (1995). |
| | 2) E. Simo-Alfonso *et al.*: *Electrophoresis* **17**: 723–731 (1996). |

Chemical formula

$$
\begin{array}{c}
-\!\!\!\!-\text{CH}_2-\!\!\!-\text{CH}\!-\!\!\!\!-_n \\
| \\
\text{C}=\text{O} \\
| \\
\text{N}-\text{H} \\
| \\
\text{CH}_2 \\
| \\
\text{CH}_2 \\
| \\
\text{O} \\
| \\
\text{CH}_2 \\
| \\
\text{CH}_2 \\
| \\
\text{OH}
\end{array}
$$

| | |
|---|---|
| Name | Poly(acryloxypropane sulfonic acid) |
| Classification• network | Natural, organic, vinyl polymer |
| Classification• crosslink point | Physical gel |
| Classification• fluid | Organogel |
| Manufacturing method | Polymerize acryloxypropane sulfonic acid in either water or an organic solvent by free radical polymerization. |
| Solvent | Water and DMF, DMSO, organic solvents like methanol and ethanol |
| Characteristics | Gel is highly absorbent |
| Gel preparation method | $\gamma$-ray, electron beam and ultraviolet light |
| Uses | Salt resistant, waterproofing materials used in civil engineering, light-resistant water-absorbing material. |
| Raw material manufacturer | |
| Product manufacturer | |
| Related literature | 1) Japanese Unexamined Patent Application publication 61-36309. |
| | 2) Japanese Unexamined Patent Application publication 57-34101 |

Chemical formula

$$
\begin{array}{c}
-\!\!\!-\text{CH}_2-\text{CH}\!-\!\!\!\!-_n \\
| \\
\text{COO}-(\text{CH}_2)_3-\text{SO}_3\text{H}
\end{array}
$$

| | |
|---|---|
| Name | Poly(acrylonitrile) [PAN] |
| Classification• network | Natural, organic, vinyl polymer |
| Classification• crosslink point | Physical gel |
| Classification• fluid | Organogel |
| Manufacturing method | Add hydrogen cyanate to acetylene under the existence of copper chloride (I) catalyst. |
| | Do a dehydration reaction after synthesizing ethylene cyanhydrin from ethylene oxide and hydrogen cyanate. |
| | Do a dehydration reaction after synthesizing lactonitrile from acetoaldehyde and hydrogen cyanate. |

Synthesize from propylene, ammonia, and air under the existence of metallic catalysts (Sohio process).

| | |
|---|---|
| Solvent | Polar organic solvent |
| Characteristics | Acrylonitrile is among the vinyl monomers that readily polymerize. It polymerizes to poly(acrylonitrile) (PAN) by a small amount of an initiator via free radical or anionic polymerization. The polymer is a white-pale yellow powder. Due to the high polarity of the side chain, nitrile group, it has strong intermolecular force, high softening point, great mechanical strength, and poor solubility. It is insoluble to ordinary organic solvents. However, it dissolves in dimethyl formamide or DMSO. |
| Gel preparation method | Dissolve electrolyte ($LiClO_4$) and polyacrylonitrile in organic solvents (propylene carbonate, dimethylformamide) and mix. After defoaming the mixed materials under reduced pressure, and when the organic solvent gradually evaporates at temperatures below the boiling point in a nitrogen atmosphere, it forms a gel material. |
| Uses | This polymer forms one of the most exceptional fibers of the vinyl polymers and is widely used in clothing. Solid-state polymeric electrolyte |
| Raw material manufacturer | Monomer: Wako Pure Chemicals, Kishida Chemicals, Tokyo Chemicals, Kanto Chemicals, Nakarai, Aldrich, etc. |
| Product manufacturer | Polymer: Wako Pure Chemicals, Aldrich, etc. |
| Special note | Acrylonitrile is an extremely poisonous liquid and will explode when mixed with air (3–17%). |
| Related literature | 1) Unexamined Utility Model Application publication (B2) 61-23947. |
| Chemical formula | |

$$\left[ CH_2 - CH \right]_n$$
$$\;\;\;\;\;\;\;\;\; |$$
$$\;\;\;\;\;\;\;\; C\;N$$

| | |
|---|---|
| Name | Polyamidine |
| Classification• network | Synthetic, organic, vinyl polymer (modified process) |
| Classification• crosslink point | Physical gel |
| Classification• fluid | Hydrogel |
| Manufacturing method | After copolymerizing acrylonitrile or methacrylonitrile and N-vinyl amides, acrylamide or methacrylamide, the poly-N-vinylamides will be hydrolyzed by the acid or base group. When it is polyarylamide or polymethacrylamide, then the polymer is base-modified by the Hoffman reaction. |
| Solvent | Water |
| Characteristics | The degree of ionic dissociation from neutral to weak base groups is rather high. A high molecular weight polymer can be formed. |

Gel preparation method

When polyamidine is included in organic wastewater and sludge, the cationic polyamidine adsorbs directly to the suspended particles ionically. This causes aggregation and gelation (a liberal interpretation).

Uses

Sludge dehydrating agent

Raw material manufacturer

Product manufacturer

Related literature

1) Mitsubishi Chemical: Japanese Unexamined Patent Application publication 6-218400.

Chemical formula

$$
\begin{array}{c}
\quad\quad\,\,\,\,/ CH_2 \,\backslash \\
+CH_2 - CR_1 \quad\quad CR_2 +_n \quad \text{or} \\
\quad\quad\backslash\,\,\,\,/ \\
\quad\quad C = N \\
\quad\quad / \\
\quad\quad N^+H_3X^-
\end{array}
\quad
\begin{array}{c}
\quad\quad\,\,\,\,/ CH_2 \,\backslash \\
+CH_2 - CR_2 \quad\quad CR_1 + \\
\quad\quad\backslash\,\,\,\,/ \\
\quad\quad N = C \\
\quad\quad\quad\backslash \\
\quad\quad\quad N^+H_3X^-
\end{array}
$$

$R_1, R_2$ : hydrogen atom or methyl group

$X^-$ negative ion

---

Name

Poly(isobutylene-co-maleic acid)

Classification• network

Synthetic, organic, vinyl polymer

Classification• crosslink point

Chemical gel

Classification• fluid

Hydrogel, organogel

Manufacturing method

Solvent

Solvent water, general organic solvent

Characteristics

Gel is relatively stable in humidity and has exceptional stability for absorption and release.

Gel preparation method

Add 1,3-dichloroisopropanol, sodium lauryl sulfonic acid 2% solution to a solution of 2-sodium salt of polyisobutylene/maleic acid anhydride and let it be keep undisturbed. After it has dried, it gels when heated for one hour at 100°C. Or, add polyethylene imine with sodium hydroxide and water to polyisobutylene/maleic anhydride copolymer, mix this well, and heat for 3 h at 160°C. Copolymer having intrinsic viscosity $[\eta] = 0.1$–8 dl/g is synthesized from ethyleneglycol, propyleneglycol (multifunctional hydroxyl groups), ethyleneglycol diglycidylether (multifunctional epoxy), and a crosslinking agent that possesses multifunctional amines.

Uses

Sanitary products, communication cables, agent to protect electrical wires from water, agent to prevent water damage in civil engineering construction, antifrost agent

Raw material manufacturer

Sales: Aldrich

Product manufacture

Kraray Isoprene (KI gel)

Special note

Shows an exceptional absorptivity for salt solutions, urine, and blood. Absorption of water from humid air when there is no actual rain is quite small, and thus it does not have to be protected. It is stable even when stored for long periods of time, even when water has been absorbed, and is durable.

Related literature

1) Unexamined Utility Model Application publication 4-41522.
2) Unexamined Utility Model Application publication 7-292023

Chemical formula

$$-(CH_2-\underset{\underset{CH_3}{|}}{\overset{\overset{CH_3}{|}}{C}}-CH_3-/-\underset{\underset{COOH}{|}}{CH}\quad\underset{\underset{COOH}{|}}{CH})_n$$

| | |
|---|---|
| Name | Poly(N-isopropyl acrylamide) [NIPAAm] [PNIPAAm] |
| Classification• network | Synthetic, organic, vinyl polymer |
| Classification• crosslink point | Chemical gel |
| Classification• fluid | Hydrogel, organogel |
| Manufacturing method | Obtain a polymer by radically polymerizing N-isopropylacrylamide in an organic solvent. In water, obtain the polymer by redox polymerization. |
| Solvent | Water and normal organic solvents (for water, the solubility is highest around 32°C) |
| Characteristics | Monomer is a white flake. |
| | The polymer has lower-limit critical solubility temperature (LCST) in water, and dissolves in water at low temperatures. It does not dissolve >32°C. |
| | At this temperature, a drastic volumetric phase transition can be seen for this gel. |
| Gel preparation method | Gel is obtained in water or organic solvent by radically polymerizing under the existence of a crosslinking agent. |
| | When doing this in water, the gel will have a volumetric phase transition if temperature is >32°C, so polymerize it at a low temperature using a polymerization accelerator. |
| Uses | Medicinal release carriers that use volume change due to the temperature, enzyme-fixing carrier, thermoresponsive absorbents. Concentration of dehydrating agents and useful materials |
| Raw material manufacturer | Kyojin |
| Product manufacturer | |
| Special note | LCST will decrease by adding inorganic salt, and will increase with surface active agents. |
| | By copolymerizing, it is possible to change LCST. |
| | Monomer has sudden oral toxicity. LD50 = 383 mg/kg (rat). |
| Related literature | 1) H. G. Schild: *Prog. Polym. Sci.* **17**: 163–249 (1992). |
| Chemical formula | |

$$CH_2=\underset{\underset{\underset{\underset{O}{\|}}{C-NH-\underset{\underset{CH_3}{|}}{CH}-CH_3}}{|}}{CH}$$

| | |
|---|---|
| Name | Polyurethane |
| Classification• network | Synthetic, organic, polyaddition polymerization |
| Classification• crosslink point | Chemical gel |
| Classification• fluid | Organogel |
| Manufacturing method | Synthesize by a polyaddition reaction of diisocyanate and multifunctional alcohol. |
| Solvent | Concentrated sulfuric acid, formic acid, phenol, cresol |

| | |
|---|---|
| Characteristics | The polymer has different properties based on R and R' types and lengths.<br><br>It is possible to create a junction zone structure by using polyether and polyester as HO–R'–OH and processing with aromatic diamine and multifunctional alcohol. This shows a rubber elasticity. It has good abrasion, acid, and oil resistance. Absorption of humidity is less than for polyamide and at a relative humidity of 65%, it is 1–1.5%. |
| Gel preparation method | Add the certain compound ratio of 4,4-diphenylmethane di-isocyanate to polyol, react it at 65°C for 60 min, and then a NCO group dual end prepolymer is obtained. In addition, a DMF solution of a chain extender is added as a chain-lengthening agent and it is reacted at 65°C. The consistent material is deaerated and then cast into a die, and it is dried under decreasing pressure until it becomes a constant temperature. |
| Uses | As mechanical properties are excellent and elasticity is high, it is used in industry as a rubber for automotive parts. Electrical insulation ability is also excellent, so it is used to make various electrical insulation parts by injection molding. It is also used as an adhesive, textile, and paint. |
| Raw material manufacturer<br>Product manufacturer | Nakarai Tesque |
| Related literature | 1) T. Terada, K. Hiraoka, and T. Yokoyama: *Japan Rubber Association Publication* **65**: 4 (1992). |

Chemical formula

$$\left[ \begin{array}{c} C-N-R-N-C-O-R'-O \\ \underset{O}{\|} \ \ \underset{H}{\,} \ \ \ \ \ \ \underset{H}{\,} \ \ \underset{O}{\|} \end{array} \right]_n$$

---

| | |
|---|---|
| Name | Polyethylene |
| Classification• network | Synthetic, organic, vinyl polymer |
| Classification• crosslink point | Chemical gel |
| Classification• fluid | |
| Manufacturing method | Industrially produced using high-temperature cracking of the distillation fraction of oil. The following polymerization methods are used.<br>1) High-pressure method—1000–2000 atm at 150–200°C with very small amount of an organic peroxide as a catalyst.<br>2) Medium-pressure method: 20–110 atm, 150–260°C, and catalyst is $Cr_2O_3 + SiO_2 - Al_2O_3 / MoO_2 - Al_2O_3$.<br>3) Low-pressure method—normal pressure, 60–80°C and $Al(C_2H_5)_3 + TiCl_4$ is catalyst. |
| Solvent | |

| | |
|---|---|
| Characteristics | A flammable gas with no color, but with an odor of olefin. Freezing point is $-169.15°C$, boiling point is $-103.71°C$, $d$ ($0°C$, $760\,mm$, air $= 1$) 0.975, explosion lower limit 3.1%, explosion upper limit 32%. Polyethylene is a clear or semiclear combustible waxy solid and has exceptional acid, alkali, water, and solvent resistance, electrical insulation, humidity prevention, and cold resistance. Thermal plasticity is good, and specific gravity is low for a synthetic resin. Mechanical strength is good and is easy to process. However, it is flammable, the thermal expansion coefficient is large, and it allows gas to pass through. The solvent affinity is rather low and the crosslinking density is high. |
| Gel preparation method | Copolymerized with divinyl compounds. Crosslinking by $\gamma$-ray irradiation in water or gelation of a surface in contact with plasma generated by glow discharge. |
| Uses | Maintaining food freshness, agricultural film, daily necessities, stationery, containers, bottles, electrical equipment |
| Raw material manufacturer | DuPont, Spencer Chemical, US Industrial |
| Product manufacturer | |
| Related literature | |
| Chemical formula | |

$$\text{+CH}_2-\text{CH}_2\,\text{+}_{\overline{n}}$$

---

| | |
|---|---|
| Name | Poly(ethylene imine) |
| Classification• network | Synthetic, organic |
| Classification• crosslink point | Chemical gel |
| Classification• fluid | Hydrogel, organogel |
| Manufacturing method | The polymer can be made into a branch structure by ring-opening polymerizing ethylene imine under an acidic catalyst. Also, a linear polymer can be made by hydrolyzing the poly(N-formylethylene imine) obtained by the ring-opening polymerization of 2-oxazoline with boron trifluoride or sulfuric acid. |
| Solvent | Water and alcohol |
| Characteristics | Shows properties of secondary amine. As it includes nitrogen atoms, hydrophilicity is high and it is cationic in water. Thus, the surface electrical charge of the suspended products and the colloids can be neutralized. There are also adsorption properties. The polymer has carcinogenic activity. |
| Gel preparation method | Gelation occurs from copolymerization of ethylene imine-ethylene oxide or the homopolymerization of polyethylene imine. Gels can be obtained by using ethyleneglycol diglycidilether, diepoxy, polyepoxy compounds as crosslinking agents. |
| Uses | Wet paper strengthener, water filtration during papermaking process, improving organic efficiency, dye-fixation agent for textiles, wastewater processing agent, ion exchange resin, enzyme stabilizing carrier. |

Raw material manufacturer    Japan Catalyst, Wako Pure Chemical Industries.

Product manufacturer

Special note    Monomer is extremely toxic. It invades the respiratory organs and kidneys and causes headaches and nausea. When inhaled for a long period of time, it decreases leukocytes

Related literature    1) Japanese Unexamined Patent Application publication 6-248073.

Chemical formula

$$+CH_2-CH_2-NH+_n$$

---

Name    Poly(ethylene oxide-*co*-propylene oxide)

Classification• network    Synthetic, organic, ring-opening polymer

Classification• crosslink point    Chemical gel

Classification• fluid    Hydrogel

Manufacturing method    Obtained by ring-opening copolymerization of ethylene oxide: propylene oxide = 50 : 50–75 : 25 (weight ratio) (molecular weight is <10,000).

Solvent    Water

Characteristics    When the molecular weight of the copolymer is low (<10,000) it does not gel by $\gamma$-ray irradiation, but it will gel if it is over 10,000 Daltons. It is difficult to dry these gels without thermal degradation. When this gel is prepared by adding poly(vinyl alcohol), it creates a superabsorbent gel.

Gel preparation method    It is possible to obtain this gel by $\gamma$-ray irradiation of this copolymer solution, but it will not gel at low molecular weight. Thus, it should be irradiated with $\gamma$-ray by adding polyvinyl alcohol to the copolymer solution.

Uses    Sanitary products

Raw material manufacturer    Japan Catalyst

Product manufacturer

Related literature    1) Japanese Unexamined Patent Application publication 58-1746

Chemical formula

$$+CH_2CH_2O-/-\overset{\overset{\displaystyle CH_3}{|}}{CH_2CHO}+_n$$

---

Name    Poly(ethylene oxide) [PEO], poly(ethylene glycol) [PEG]

Classification• network    Synthetic, organic, ring-opening polymer

Classification• crosslink point    Chemical gel

Classification• fluid    Hydrogel, organogel

Manufacturing method    Obtained by the ethylene oxide ring-opening polymerization using Friedel–Crafts catalysts such as boron trifluoride, tin tetrachloride or zinc chloride, or acid or alkali catalysts.

Solvent    General organic solvents

Characteristics

Based on the polymerization conditions, products that range in molecular weight from 200 to over 1,000,000 can be obtained. Depending on the molecular weight, a range from liquids to hard solids (such as waxes) can be obtained. It has good chemical resistance, and bending characteristics, and those with high molecular weight have high crystallinity. Tensile strength is also high and it does not absorb humidity. Toxicity is extremely low, with low immunogenicity, and thus it is useful as a biocompatible material.

Gel preparation method

React ethylene glycol using triisocyanate as a crosslinking agent and triethylamine as a catalyst in dioxane. Normally, reaction temperature is 60°C. $\gamma$-ray is irradiated to the polymer solution.

Uses

Surface active agents, neutral detergents, medicine, perfumes, plastic materials, thickeners

Raw material manufacturer

Seitetsu Chemical, Meisei Chemical

Product manufacturer

Kishida Chemical, others

Related literature

1) Y. Gnaunou, G. Hild, and R. Pempp: *Macromolecules* (1984).
2) Chemistry

Chemical formula

$HO+CH_2CH_2O+_nCH_2CH_2OH$

---

Name

Poly(ethylene-*co*-vinyl acetate) [EVA]

Classification• network

Synthetic, organic, vinyl polymer

Classification• crosslink point

Chemical gel

Classification• fluid

Organogel

Manufacturing method

Obtained by polymerizing using solvents like tert-butanol under the existence of a free radical initiator at 300–1000 kg/cm$^2$. The industrial manufacturing methods used for EVA include: vinyl acetate (VA) concentration into the high-pressure free radical polymerization (VA concentration 0–45%); solution polymerization (VA concentration 45–70%); and emulsion polymerization (VA concentration 70–100%). Generally, EVA resin is manufactured using a high-pressure free radical polymerization method. This method compresses ethylene and vinyl acetate monomers under high pressure (1000–2000 kg/cm$^2$), and it is then polymerized under high temperature in a reactor while injecting the catalysts. After this, it is separated from the residual monomer and is bagged after it has been extruded into pellets.

Solvent

Chloroform, benzene, etc.

| | |
|---|---|
| Characteristics | Thermoplastic. Some of these copolymers are flexible without adding a plasticizer. For certain applications, they are superior to silicone rubber. They are stable and can be readily sterilized. DDS Ocusert of Alza uses EVA copolymer film and Progestasert uses EVA polymer film. These copolymers with low vinyl acetate contents have greater hardness and have excellent mechanical and physical properties (used for improving wax or as adhesives). Those with high vinyl acetate content have the properties of rubber. This rubber has excellent compatibility with other rubbers and has good rolling characteristics and processability. They can be crosslinked by peroxides. They lack resistance to low temperature and solvents. |
| Gel preparation method | Free radical polymerization in the presence of divinyl monomer |
| Uses | Release control film for drugs, foam (lumber), film (agriculture, food wrapping), various polymer improvers |
| Raw material manufacturer | Ocusert (intraocular treatment system), Progestasert (intrauterine contraceptive system), transderm-Nitro (endodermic treatment system) |
| Product manufacturer | Wako Pure Chemical Industries |
| Related literature | 1) Organic Synthetic Chemistry Association (ed.): *Organic Compounds Dictionary*, Kodansha Scientific. |
| | 2) Polymer Dictionary Editorial Association (ed.): *Polymer Dictionary*, Asakura Shoten. |
| Chemical formula | |

$$-\!\!\left[CH_2 - CH_2\right]_m\!\!\left[CH_2 - CH\right]_n\!\!-$$
$$\underset{\displaystyle OCOCH_3}{|}$$

| | |
|---|---|
| Name | Poly(organosiloxane) |
| Classification• network | Synthetic, organic, polycondensation |
| Classification• crosslink point | Chemical gel |
| Classification• fluid | Organogel |
| Manufacturing method | Generally obtained by polymerizing organic silicone compounds at the same time as hydrolysis is done (siloxane). |
| Solvent | General organic solvents |
| Characteristics | |
| Gel preparation method | When a substituent on the silicon is replaced with organic groups like hydrogen, methyl or phenyl groups, it is called organosilane. Polymers that are long, such as –Si–O–Si–O, are called poly(organosiloxane). |
| Uses | Lubricants, resin, rubber, humidity sensors |
| Raw material manufacturer | Aldrich, Sin-Etsu Chemical Co, Wako Pure Chemical Industries, Tokyo Chemical, others (tetra ethoxysilane) |
| Product manufacturer | Shin-Etsu Chemical Co, Toshiba Silicone, Dow Corning, Hexsite, others |
| Special note | Since alkoxysilane has higher reactivity, it may hydrolyze |

| Related literature | 1) Y. Sakai, Y. Sadaoka, M. Matsuguchi, N. Moriga, and M. Shimada: Humidity sensors based on organopolysiloxane having hydrophilic groups. *Sensors and Actuators* **16**: 359–367 (1989). |
|---|---|
| Chemical formula | |

$$R \longrightarrow \left[ \underset{\displaystyle R}{\overset{\displaystyle R}{Si}} \longrightarrow O \right]_{n} R$$

---

| Name | Poly(chloromethyl styrene) |
|---|---|
| Classification• network | Synthetic, organic, vinyl polymer |
| Classification• crosslink point | Chemical gel |
| Classification• fluid | Organogel |
| Manufacturing method | |
| Solvent | General organic solvents |
| Characteristics | Raw material monomer: para isomer is bp 229°C, refractive index is $n_D^{20} = 1.5740$, specific gravity 1.083, flashpoint 104°C. |
| Gel preparation method | After polymerizing, react with N,N,N,N'-tetramethylhexanediamine and crosslink and quaternize at the same time. |
| Uses | Raw material comonomer for ion exchange resin, comonomer for functional polymer, humidity sensor. |
| Raw material manufacturer | Aldrich, Acros Organics, Tokyo Chemical, Kishida Chemical |
| Product manufacturer | |
| Special note | It is also called vinyl benzylchloride. Reacts easily in air and light, and with No. 3 oils. |
| Related literature | 1) Y. Sakai, Y. Sadaoka, M. Matsuguchi, and H. Sakai: Humidity sensor durable at high humidity using simultaneously cross-linked and quaternized poly(chloromethyl styrene). *Sensors and Actuators* B **24**: 689–691 (1995). |
| Chemical formula | |

$$\left[ CH_2 - CH \right]_{n}$$

with benzene ring bearing $-CH_2Cl$

---

| Name | Poly(dioxolan) |
|---|---|
| Classification• network | Synthetic, organic, ring-opening polymer |
| Classification• crosslink point | Chemical gel |
| Classification• fluid | Hydrogel |
| Manufacturing method | Obtained by the cationic polymerization of 1,3-dioxolan, 2-methyl-1,3-dioxolan, 4-methyl-1,3-dioxolan, 2,2'-diethyl-1,3-dioxolan. Or it can be obtained from the polycondensation of ethylene glycol and formaldehyde. |

| | |
|---|---|
| Solvent | Water |
| Characteristics | Hydrophilic polymer |
| Gel preparation method | Gel is created from polymer solution using $\gamma$-ray irradiation or a crosslinking agent having multifunctional reaction groups (isocyanate and carboxylic acid). In order to obtain a homogeneous gel. $\gamma$-ray irradiation is preferred. |
| Uses | Absorbent materials |
| Raw material manufacturer | Kanto Chemical, Wako Pure Chemical Industries |
| Product manufacturer | |
| Related literature | 1) Japanese Unexamined Patent Application publication 8-20640 |
| Chemical formula | |

$$\{CH_2O - CH_2CH_2O\}_n$$

| | |
|---|---|
| Name | Poly(dimethylaminopropyl acrylamide) [PDMAPAAm] |
| Classification• network | Synthetic, organic, vinyl polymer |
| Classification• crosslink point | Chemical gel |
| Classification• fluid | Hydrogel, organogel |
| Manufacturing method | It is easily solution, emulsion, suspension and bulk polymerized using a normal polymer initiator using dimethylaminopropylacrylamide as the raw material. Heat of polymerization is 16 kcal/mol. |
| Solvent | Water and normal organic solvents. Dissolves in water at various concentrations. |
| Characteristics | Dimethylaminopropylacrylamide will not hydrolyze even if left for 24 h at 100°C at a pH of 3–12. It shows a strong basicity compared to dimethylaminoethyl (meth)acrylate. |
| Gel preparation method | Gels in 18–24 h at 45–60°C when using potassium persulfate as the polymerization initiator and N,N'-methylene-bis-acrylamide as a crosslinking agent. |
| Uses | Poly(dimethylaminopropylacrylamide) is used as a cationic polymer flocking agent and dispersion agent for resins that utilizes the characteristics of the amino group. It is also used to formulate chemicals for paper-strengthening agents. |
| Raw material manufacturer | Kyojin |
| Product manufacturer | Kojin Co. Ltd DMAPAA |
| Special note | Dimethylaminopropylacrylamide shows oral toxicity in mice at $LD_{50} = 5300$ mg/kg. It causes a reaction to skin and mucous membranes. |
| Related literature | 1) Kojin Ltd DMAPAA. |
| Chemical formula | |

$$\{CH_2 - CH\}_n$$
$$C - NH - CH_2 - CH_2 - CH_2 - N\{CH_3\}_2$$
$$\| $$
$$O$$

| | |
|---|---|
| Name | Poly(styrene sulfonic acid) |
| Classification• network | Synthetic, organic, vinyl polymer |
| Classification• crosslink point | Chemical gel |

Classification• fluid — Hydrogel

Manufacturing method — Sulfonate by adding polystyrene which was polymerized by AIBN and dissolved in cyclohexane into sulfuric acid in which $P_2O_5$ was dissolved.

Solvent — Water

Characteristics — Strong acid. Stable to organic solvents, strong acids, and strong alkali.

Gel preparation method — Copolymerized with styrene in the presence of divinyl benzene that is a crosslinking agent.

Uses — Cationic exchange resin.
Softening of hard water, recovery of metallic ions, refining of materials, dispersion agents, antistatic agent.

Raw material manufacturer — Nakarai Tesque (as Na sodium)

Product manufacturer — Mitsubishi Chemical

Special note — When the sulfonic acid group is an H-type, it decomposes easily in the solid state.

Related literature —
1) H. Vink: *Makromol. Chem.* **182**: 279 (1981).
2) M. P. Tsyurupa, V. A. Davankov, and S. V. Rogozhin: *J. Polym. Sci. Symp.* **47**: 189 (1974).

Chemical formula —

$$+CH_2-CH+_n$$

$SO_3H$

---

Name — Poly(styrene-*co*-butadiene)

Classification• network — Synthetic, organic, vinyl polymer

Classification• crosslink point — Physical gel

Classification• fluid — Organogel

Manufacturing method — Obtained by adding styrene and butadiene gradually using sodium naphthalene and organic lithium as initiators and then live anionic polymers. By coupling the living polymers using coupling agents such as carbonyl chloride and thionyl chloride, branched block copolymers can be produced.

Solvent — Aromatic hydrocarbon, halogen hydrocarbon

Characteristics — General characteristics: At room temperature, the polymer shows vulcanized rubber properties and exhibits thermoplasticity at elevated temperatures. It shows resin properties following increase in styrene concentration. When using trans-polybutadiene, it shows a shape memory tendency. It is possible to process while melted and is strong and has high-elasticity and good low-temperature characteristics. It also has exceptional acid resistance and alkali resistance.

Gel characteristics: The polymer blend material allows gels with various properties to be designed by changing the low molecular weight material.

| | |
|---|---|
| Gel preparation method | Gel-like polymer blends are created when a styrene-butadiene block copolymer (polymer organic material) unit with a number-average molecular weight >20,000 and low molecule organic material with a number-average molecular weight <20,000 are mixed using a high-speed blender (>300 rpm) so that the ratio of the volume rate of the polymer organic material is <30%. |
| Uses | Hot melt-type adhesives, sealant, caulking, wiping material, injection-molded parts for automobiles, plastic modifier, electrical cable covers. |
| Raw material manufacturer | Aldrich |
| Product manufacturer | |
| Special note | Styrene and butadiene monomers will stimulate the skin, eyes, and nasal mucous membranes and cause inflammation. It also can cause paralysis. |
| Related literature | 1) Y. Furubori: Japanese Unexamined Patent Application publication 5-239256. |
| | 2) T. Shiroki: Japanese Unexamined Patent Application publication 57-49645. |
| | 3) T. Shinozaki: *Nikkei New Material* **54**: 40 (1988). |
| Chemical formula | |

$$\pm CH - CH_2 \pm_m \ / \ \pm CH_2 - CH = CH - CH_2 \pm_n$$

| | |
|---|---|
| Name | Polystyrene (styrol, vinylbenzene, phenylethylene) |
| Classification• network | Synthetic, organic, vinyl polymer |
| Classification• crosslink point | Chemical gel |
| Classification• fluid | Organogel |
| Manufacturing method | There are various chemical manufacturing methods for monomers. A representative example of this would be the dehydrogenation of ethylbenzene. This can be done by passing ethylbenzene with steam through such acidic-mixed catalysts as Zn, Cr, Ca and Mg, or other dehydrogenation catalysts at 600°C. The reaction yield is 30–40%, with total yield of 85–90%. |
| Solvent | The monomer dissolves well in organic solvents such as methanol, ethanol, and acetone. |
| Characteristics | The polymer dissolves well in methanol, ethanol, and acetone. The typical physical properties of styrene include: molecular weight 104.14; specific gravity (25°C) 0.9019; refractive index (25°C) 1.5439; boiling point 145.2°C; specific heat (25°C) 0.416; heat of polymerization 160.2 cal/g (properties of styrene and polystyrene are described in detail in the monograph [1] listed here under related literature. |
| Gel preparation method | Polystyrene is a clear, thermoplastic resin. Under the existence of divinyl monomers as a crosslinking agent, a copolymer gel is obtained by copolymerization with other vinyl monomers (sodium styrene sulfonic acid). When the feed ratio of styrene is high, the gel experiences difficulty if it includes solvent. |

| | |
|---|---|
| Uses | It is important industrially as a raw material for individual or specific polymers. It becomes the manufacturing raw material for styrene butadiene rubber and polystyrene. Paints, polyester resins, and ion exchange resins are also made. |
| Raw material manufacturer | Junsei Chemical, others |
| Product manufacturer | Junsei Chemical (product number 39380-1235), others |
| Special note | |
| Related literature | 1) T. Fujii and I. Sumoto: Plastic Materials Seminar 12, Styrol Resins (Nikkan Kogyo Shimbunsha). |
| | 2) *Chemistry Dictionary*, Kyoritsu Publ. |
| | 3) *Polymer Dictionary*, Asakura Shoten. |
| Chemical formula | |

$$+CH_2-CH+_n$$

---

| | |
|---|---|
| Name | Poly(tetrafluoroethylene) |
| Classification• network | Synthetic, organic, vinyl polymer |
| Classification• crosslink point | Chemical gel |
| Classification• fluid | |
| Manufacturing method | Monomer is obtained by causing a dehydrochlorination when Freon-22 ($CClF_2$) is sent through a tube that has platinum, silver, or carbon heated to 650–800°C (90% yield). It can also be obtained by thermal decomposition of the sodium salt of trifluoroacetic acid in the presence of sodium hydroxide. |
| Solvent | |
| Characteristics | Monomer is a gas with no color or odor and burns in air. The solidification point is 142.5°C, boiling point is 76.3°C, critical pressure is 40.2 kg/cm², and critical density is 0.58 g/ml. Polymer is a white or light-gray solid with $d = 2.1$–2.3, $n_D$ 1.775, and the melting point 327°C. There is no water absorption and light waves of 200–400 nm pass through it. It is not flammable, it is acid-resistant, alkali-resistant, water-repellent, oil-repellent, has exceptional electrical characteristics, and is stable for all organic solvents. It starts to decompose at 400°C, and completely decomposes at 600–700°C. Solvent affinity is low and the crosslinking density is high. |
| Gel preparation method | Surface is crosslinked and gelled by contacting an inactive gas that was excited, using high-frequency electrical waves on the surface of the polymer using the radical polymerization method (suspension polymerization or emulsification polymerization method). |
| Uses | Machine parts such as packing, tubes, sheets, gears, and bearings. Coverings for electrical wires, electrical equipment parts, lining, rolls, and chemical test equipment. |
| Raw material manufacturer | Osaka Metal Industries KK (Polyflon)/DuPont (teflon), etc. |
| Product manufacturer | |
| Related literature | |

Chemical formula

$$-\!\!+\!CF_2-CF_2\!+\!\!_n\!-$$

| | |
|---|---|
| Name | Poly(2-hydroxyethyl methacrylate) [HEMA, PHEMA] |
| Classification• network | Synthetic, organic, vinyl polymer |
| Classification• crosslink point | Chemical gel |
| Classification• fluid | Hydrogel, organogel |
| Manufacturing method | Monomer is created by addition reaction of ethylene oxide and methacrylic acid (MMA). |
| Solvent | Water, organic solvents (slight dissolving). |
| Characteristics | Divinyl monomer is contained as an impurity, so under normal refining methods, it naturally becomes a gel by free radical polymerization even without crosslinking agents. The moisture content of gel is around 45–50 wt.%, and is very soft, so it is a good material to be used for contact lenses. |
| Gel preparation method | By heating and polymerizing monomer solution at several tens of °C for several hours, a polymer is formed that has 3D crosslinking. By boiling this in saline solution after removal from the mold, solvent and residual monomer are extracted and removed. Finally, a highly hydrated hydrogel is prepared by hydration. |
| Uses | For contact lenses |
| Raw material manufacturer | Wako Pure Chemical Industries, Junsei Chemical, Aldrich, others |
| Product manufacturer | |
| Special note | Contact lenses available on the market (HEMA materials) Product name (manufacturing company) Zero 4 (American Hydron), Hydrocurvell$_{55}$ (Barnes-Hind), O$_4$ (Bausch & Lomb), Softcon EW (CIBA Vision Care), Parmalens (Cooper Vision), Vistamare (Vistakon), Durasoft 3 (Welsey-Jessen) |
| Related literature | 1) *Experimental Cell Research* **143**: 15–25 (1983). 2) Association of Organic Synthesis Chemistry (ed.): *Organic Compound Dictionary*, Kodansha Scientific. |

Chemical formula

| | |
|---|---|
| Name | Poly(vinylalcohol) [PVA] |
| Classification• network | Synthetic, organic, vinyl polymer |
| Classification• crosslink point | Physical gel, chemical gel |
| Classification• fluid | Hydrogel |
| Manufacturing method | Hydrolysis of poly(vinyl acetate) by alkali or acid. |
| Solvent | Water |
| Characteristics | The appearance of polymer is a white powder, density is 1.21 ~ 1.31 g/cm³, and secondary transition point of 65–85°C. |
| Gel preparation method | Freeze-dried dehydrated PVA that was completely saponified, with repeated freezing and defrosting of PVA solution, chemically reacting with N-methylol compounds, dicarboxylic acid and bis epoxy, adding metals such as steel and copper, or irradiating radiation. |

It is also obtained by irradiating the PVA which was acetallated with formylstyryl pyridinium salt.

Uses                          Artificial joints, artificial glass
Raw material manufacturer     Clare, Japan Synthetic Chemicals
Product manufacturer
Related literature            1) M. Watase, K. Nishinari, and M. Nanbu: *Polym. Commun.* **24**: 52 (1983)
                              2) M. Watase: *Japan Chemistry Association Publication* **973**(7), (1983).
                              3) K. Ichimura and S. Watanabe: *J. Polym. Sci., Polym. Chem. Ed.* **20**: 1419 (1982).

Chemical formula

$$\left. +CH_2-\underset{\underset{OH}{|}}{CH} \right\}_n$$

---

Name                              Poly(vinylpyridine)
Classification• network           Synthetic, organic, vinyl polymer
Classification• crosslink point   Chemical gel
Classification• fluid             Organogel, hydrogel
Manufacturing method              Radical or anionic polymerization using pyridine with an initiator.
Solvent                           Dissolves in ether: difficult to dissolve in water.
Characteristics                   As it reacts with various nucleophilic reagents, it is used as an organic intermediate.
                                  2-vinyl pyridine: an oil-based liquid that polymerizes easily. Boiling point is 79–82°C/29 mm. It solidifies when heated to 158–160°C at ambient pressure.
                                  3-vinyl pyridine: an unstable, yellowish liquid with a boiling point of 67–68°C/13 mm. Polymerizes at room temperature.
                                  4-vinyl pyridine: a liquid with no color and irritating odor. Boiling point is 65°C/15 mm, and polymerizes or dissociates rapidly with sunlight.
Gel preparation method            Free radical or anionic polymerization is possible and can gel using a crosslinking agent (N,N′-methylene bis acrylamide, divinyl benzene, etc.) that has a multifunctional group. Various copolymerized crosslinked materials can be obtained by comonomers.
Uses                              Special synthetic rubber, adhesive for tire cords, polymer detergents, ion exchange resins, dye-improvers for acrylic textiles.
Raw material manufacturer         4-vinylpyridine: Koei Chemical
Product manufacturer
Special Items                     According to fire safety law, it is a toxic material. Depending on the position on the pyridine ring, three isomers exist (2-vinyl, 3-vinyl or 4-vinyl pyridine). The 2-vinyl pyridine is synthesized by condensing α-pycoline with formaline while dehydrating during distillation. A 3-vinyl pyridine is synthesized by reducing 3-acetyl pyridine and dehydrating. A 4-vinyl pyridine is synthesized by condensing γ-pycoline with formaline. A 2- and a 4-vinyl pyridine are commercially produced.

| | |
|---|---|
| Related literature | 1) Chemistry Dictionary Editorial Association (ed.): *Chemistry Dictionary*, Kyoritsu Publ. (1963). |
| Chemical formula | |

$$\math{+CH-CH_2+}_n \qquad \math{+CH-CH_2+}_n \qquad \math{+CH-CH_2+}_n$$

(1) 2-vinyl pyridine   (2) 3-vinyl pyridine   (3) 4-vinyl pyridine

---

| | |
|---|---|
| Name | Poly(vinyl pyrrolidone) [PVP] |
| Classification• network | Synthetic, organic, vinyl polymer |
| Classification• crosslink point | Chemical gel |
| Classification• fluid | Hydrogel, organogel |
| Manufacturing method | Obtained by adding a little ammonia to vinyl pyrrolidone solution and polymerizing with a hydrogen peroxide catalyst. |
| Solvent | Water, alcohol, chloroform |
| Characteristics | Polymer is a white powder and the film is clear and brittle. Industrially, $M_w = 2.5 \times 10^4$–$1.2 \times 10^6$ type is manufactured. |
| Gel preparation method | Copolymerize N-vinyl-2-pyrrolidone with crosslinking agent such as ethylene glycol and swell in solvents. |
| Uses | Dye-aiding agent, cosmetic additive, soft contact lenses |
| Raw material manufacturer | Sigma |
| Product manufacturer | Wako Pure Chemical Industries, Nakarai Tesque |
| Related literature | 1) F. X. Quinn, E. Kampff, G. Smyth, and V. J. McBrierty: *Macromolecules* **21**: 3191 (1988). |
| | 2) R. F. Ofstend and C. I. Poser: *ACS Polym. Mater. Sci. Eng.* **57**: 366 (1987). |
| Chemical formula | |

$$\math{+ CH_2-CH +}_n$$

---

| | |
|---|---|
| Name | Poly(vinylmethylether) [PVME] |
| Classification• network | Synthetic, organic, vinyl polymer |
| Classification• crosslink point | Chemical gel |
| Classification• fluid | Hydrogel |
| Manufacturing method | Monomer is synthesized by heating acetylene and methanol under pressure in the presence of potassium hydroxide as a catalyst. Polymer is obtained using $BF_3$, $I_2$, and $AlCl_3$ as cationic polymerization catalysts. Maintain monomer under pressure at 5°C and add dioxane solution of boron trifluoride and heat gradually to 100°C. When this is polymerized, an atactic polymer like a thick malt syrup is formed. When dried propane is used as a solvent at $-78$°C, and when boron trifluoride diethyletherate is used as a catalyst, a crystalline isotactic polymer is obtained. |
| Solvent | Water |

| | |
|---|---|
| Characteristics | Atactic polymers will dissolve in cold water, methanol, acetone and benzene, but will not dissolve in water that is hotter than 36°C. Isotactic polymers are rubbery solids that will not dissolve in water. Gels made from atactic polymers have a phase transition around 36°C and will swell and absorb water at low temperatures and will shrink and release water at high temperatures. The response speed of sponge and fibrous gels is several tens of seconds to several hundred milliseconds. A homogeneous gel would take several hundred days to produce. |
| Gel preparation method | After polymer aqueous solution is defoamed under reduced pressure, it is crosslinked by irradiating with 100 kGy of γ-rays or electron beam, and a gel is formed. |
| Uses | Atactic polymer is used for gum-label adhesives, polymer plasticizers, and textile finishing agents. |
| Raw material manufacturer | Polymer: BASF, Aldrich, Tokyo Chemical |
| Product manufacturer | Gel: None (research scale: Material Engineering, Industrial Engineering Research Center) |
| Related literature | 1) Xia Huang *et al.*: *J. Chem. Eng. Japan* **20**: 123 (1987). |
| | 2) K. Hirasa: *Polymer Theses* **45**(11): 661 (1989). |
| | 3) R. Kishi *et al.*: *J. Intelligent Mater. Syst. Struct.* **4**: 533 (1993). |
| Chemical formula | |

$$+CH-CH_2\,\tfrac{}{\,n}$$
$$\quad\ \mid$$
$$\ \ OCH_3$$

---

| | |
|---|---|
| Name | Poly(propylene oxide) |
| Classification• network | Synthetic, organic, polysaccharides |
| Classification• crosslink point | Chemical gel |
| Classification• fluid | Hydrogel, organogel |
| Manufacturing method | Propylene oxide is obtained from the ring closure of alkali of 1-chloro-2-propanol or 2-chloro-1-propanol under alkaline conditions. By ring-opening polymerization, polymer is obtained when reacting with active hydrogen compounds like alcohol, fatty acids, or amines. |
| Solvent | Water, alcohol, ether |
| Characteristics | The block copolymer with ethylene oxide becomes a polymeric surfactant. The low molecular weight polypropylene oxide is water soluble, but becomes insoluble in water when molecular weight is >900. Atactic structure is amorphous and the isotactic structure is a crystallized solid with a melting point of 70°C. |
| Gel preparation method | Gels using multifunctional crosslinking agents |
| Uses | Polyester resin raw material, urethane foam raw material, poly(vinyl chloride) stabilizer, detergents, pigment, intermediate for medical products, synthetic resin raw material, sterilizer. |
| Raw material manufacturer | Asahi Glass, Tokuyama Sosan, Showa Denko, Mitsui Toatsu Kagaku, Daicel Chemical, Japan Okisiran |
| Product manufacturer | |
| Special note | Propylene oxide has a high volatility and burns easily. When gas mixes with air, it can explode |

Related literature
Chemical formula

$$CH_3$$
$$\hspace{-CH_2CHO +_n}$$

| | |
|---|---|
| Name | Poly(perfluorocarbon sulfonic acid) |
| Classification• network | Synthetic, organic, vinyl polymer |
| Classification• crosslink point | Physical gel |
| Classification• fluid | Hydrogel |
| Manufacturing method | It is obtained by copolymerizing tetrafluoro ethylene and perfluorosulfonyl ethoxy divinylether and then hydrolyzing it |
| Solvent | Water |
| Characteristics | Polymer is acid- and alkali-resistant. |
| | Permeates alkali metallic cation selectively and does not allow a negative ion through. |
| | $T_g \approx 130°C$ |
| | As it easily hydrates, it absorbs water and swells. |
| Gel preparation method | Physically crosslinked by crystal formation of tetrafluoro ethylene skeleton. |
| Uses | Cation exchange film |
| | Electrolyte barrier |
| | Demineralization |
| Raw material manufacturer | |
| Product manufacturer | DuPont, Dow Chemical, Asahi Chemical, Asahi Glass Co. |
| Related literature | 1) Electrical Chemistry Association (ed.): *Electrical Chemistry Manual* No. 4, Maruzen (1985). |
| | 2) T. D. Gierke, G. E. Munn, and F. C. Wilson: *J. Polym. Sci. Polym. Phys. Ed.* **19**: 1687 (1981). |

Chemical formula

| | |
|---|---|
| Name | Poly(methacrylic acid-*co*-ethyleneglycol) |
| Classification• network | Natural, organic, vinyl polymer |
| Classification• crosslink point | Physical gel |
| Classification• fluid | Hydrogel, organogel |
| Manufacturing method | Ethylene glycol—Hydrolysis of ethylene dibromide under alkali conditions, oxidation of ethylene by hydrogen peroxide, hydrolysis of ethylene chlorohydrin by $Na_2CO_3$ or $NaHCO_3$ in autoclave, or hydrolysis of ethyleneoxide by dilute sulfuric acid. |
| | Methacrylic acid—Hydrolysis of methyl methacrylate by NaOH, hydrolysis of acetonecyanhydrin by sulfuric acid, oxidization of meta-acrolane or dehydration of $\alpha$-oxyisobutyric acid by $P_2O_5$. |
| Solvent | Water |
| Characteristics | Ethylene glycol—viscous liquid with no color and a sweet flavor. mp $-11.5°C$ ($-15.6°C$), bp $197°C$, $109°C/25\,mm$, $93°C/13\,mm$. It absorbs humidity. It freely mixes with EtOH, MtOH, acetone, glycerine, acetic acid, and pyridine in addition to water. Dissolves KOH, $Ca(OH)_2$, $K_2CO_3$. |

Methacrylic acid—No color and columnar crystal. Has irritating odor. mp 16°C, bp 160.5°C. Dissolves with alcohol and ether at any concentration. Does corrode.

It is easy to synthesize, but it is weak if it undergoes environmental changes and a sol-gel transition is easily caused by changes in pH and humidity; this further lowers mechanical strength.

Gel preparation method
Polymerize polyethylene glycol under existence of alkali and halogenated ethylene. When methacrylic acid is mixed after polymerization with heat and catalysts, a polymer complex is formed by hydrogen bonding (hydrophobic bonding also happens) and gel is formed.

Uses
Ethylene glycol is used for synthetic fibers, resins, dynamite, detergents, medicine, cosmetics, fragrances, various raw materials. Methacrylic acid is used as a raw material for vinyl compounds and ion exchange resins.

Raw material manufacturer
Wako Pure Chemical Industries, Tokyo Chemical

Product manufacturer

Related literature

Chemical formula

$$\begin{array}{ccc} & CH_3 & CH_3 & CH_3 \\ & | & | & | \\ -[CH_2-C\phantom{xx}]_n & CH_2-C & -[CH_2-C\phantom{xx}]_n \\ & | & | & | \\ & C=O & C=O & C=O \\ & | & | & | \\ & O-H & O-H & O-H \end{array}$$

$$-[CH_2CH_2-O]_n-CH_2CH_2-O-[CH_2CH_2-O]_n-$$

| | |
|---|---|
| Name | Poly(methylmethacrylate), poly[1-(methoxycarbonyl)-1-methylethylene] [PMMA] |
| Classification• network | Synthetic, organic, vinyl polymer |
| Classification• crosslink point | Chemical gel |
| Classification• fluid | Organogel |
| Manufacturing method | Manufacturing method for monomer: manufacturing method for methacrylic acid: [(i) cyanhydrin is made from acetone and hydrogen cyanide followed by hydrolysis by sulfuric acid; and (ii) isobutylene or t-butyl alcohol is oxidized.] These reactions are performed under methanol. Manufacturing method for polymer: free radical or anionic polymerization. |
| Solvent | 1,2-dichloroethane, ethyl acetate, acetone, toluene, acetic acid, formic acid |
| Characteristics | Polymer characteristics: it is a clear crystalline polymer. It is stable under sunlight and does not change color. Transmission of light is exceptional. The disadvantage is that it is easily scratched. It has good chemical resistance and is stable under 10% sulfuric acid, hydrochloric acid, nitric acid, sodium hydroxide, ammonia water, and saltwater at room temperature. |

Monomer characteristics: mp $-48°C$, bp $100.8°C$, $46°C$
($160\,mmHg$), $29°C$ ($50\,mmHg$), $d_4^{20} = 0.943$ $n_D^{20} = 201.416$,
flashpoint $13-30°C$ (open), $10°C$ (closed), viscosity $0.58\,cP$
($20°C$), copolymerization constant $Q$ $0.74$, e $0.40$. Easy to cause
free radical polymerization, including ultraviolet-initiated
polymerization. It also polymerizes anionically.

| | |
|---|---|
| Gel preparation method | Radical polymerization (copolymerization with divinyl compounds) |
| Uses | Speed control film, plastic glass sheets, compounding materials, paints, construction materials, cosmetic dentistry materials, optical materials. Used also as a comonomer of MBS resins. |
| Raw material manufacturer | Aldrich, Wako Pure Chemical Industries, ACROS, CHIMICA, NV, others. |
| Product manufacturer | Mitsubishi Rayon, Asahi glass |
| Special note | Monomer is poisonous, $LD_{50}$ $7.9\,g/kg$ [W. Deichman and J. Ind. Hyg.: *Toxicol.* **23**: 343 (1941).] |
| Related literature | 1) H. Ulrich: *Introduction to Industrial Polymers*, 2nd ed., Munich-Vienna-New York-Barcelona: Hanser Publishers. |
| | 2) Organic Synthetic Chemistry Association (ed.): *Organic Compound Dictionary*, Kodansha Scientific. |

Chemical formula

$$-\!\!\!+\!\!\Big[\begin{array}{c} CH_3 \\ | \\ C-CH_2 \\ | \\ COOCH_3 \end{array}\Big]_{\!\!\overline{n}}$$

---

| | |
|---|---|
| Name | Poly(N-methylolacrylamide) |
| Classification• network | Synthetic, organic, vinyl polymer |
| Classification• crosslink point | Chemical gel |
| Classification• fluid | Hydrogel, organogel |
| Manufacturing method | Heat acrylamide and p-formaldehyde with basic catalyst. In this case, water, alcohols, or halogenated carbons are used as solvents. |
| Solvent | Alcohols |
| Characteristics | Monomer: mp $75°C$, copolymerization constant $Q$ $0.31$, e $0.36$, white crystalline powder. |
| Gel preparation method | Radical copolymerization with divinyl compound by N,N,N′,N′-tetramethylhexanediamine (TEMED) as the polymerization accelerator, and ammonium peroxodisulfate as an initiator. |
| Uses | Fiber processing agent, paper processing resin, raw material for pain adhesives, comonomers for various synthetic resins to improve hydrophilicity. The gel copolymerized with acrylamide is used for electrophoresis testing (DNA sequence). |
| Raw material manufacturer | Aldrich |
| Product manufacturer | Tokyo Chemical, Kishida Chemical, Nakarai, Sankyo Chemical, Soken Chemical |
| Related literature | 1) R. J. Molinari *et al.*: *Adv. Electrophoresis* **6**: 43–60 (1993). |
| Chemical formula | |

$$+ CH_2 - CH +_{\overline{n}}$$
$$| $$
$$C = O$$
$$| $$
$$N - H$$
$$| $$
$$CH_2OH$$

| | |
|---|---|
| Name | Poly(methoxy ethyleneglycol methacrylate) |
| Classification• network | Synthetic, organic, vinyl polymer |
| Classification• crosslink point | Physical gel |
| Classification• fluid | Organogel |
| Manufacturing method | The electrical conductivity of the salt-in-polymer-type polymer electrolyte prepared by dissolving an alkaline salt such as lithium trifluoromethanesulfonic acid lithium into polymethoxy ethyleneglycol methacrylate. The electrical conductivity of the polymerized salt-in-polymer-type polymer electrolyte is a function of the combination of temperature and salt. The Arrhenius plot becomes linear. |
| Solvent | Benzene, chloroform, other |
| Characteristics | Gel is highly absorbent |
| Gel preparation method | Obtain polymer by isobisisobutyronitrile as an initiator to polymerize it in benzene at 65°C for 48 h. After refining, evaporate the solvent slowly from the mixed solution that dissolved the alkaline salt along with the lithium trifluoromethanesulfonic acid in the anhydride methanol solvent. |
| Uses | Solid-state auxiliary battery using a gel electrolyte |
| Raw material manufacturer | |
| Product manufacturer | |
| Related literature | 1) Du Wei Xia, D. Soltz, and J. Smid: Conductivities of solid polymer electrolyte complexes of alkali salts with polymers of methoxypolyethyleneglycol methacrylates. *Solid State Ionics* **14**: 221–224 (1984). |
| | 2) Du Wei Xia and J. Smid: Solid polymer electrolyte complexes of polymethacrylates carrying pendent oligo-oxyethylene (Glyme) chains, *J. Polym. Sci., Polym. Letters* **22**: 617–621 (1984). |
| Chemical formula | |

$$CH_3$$
$$|$$
$$+CH_2 - C +_{\overline{n}}$$
$$|$$
$$COO + CH_2CH_2O +_{\overline{n}} CH_3$$

| | |
|---|---|
| Name | Poly(L-leucine) • sponge (Xemex Epicuel™) |
| Classification• network | Synthetic, organic, vinyl polymer |
| Classification• crosslink point | Physical gel |
| Classification• fluid | Organogel |
| Manufacturing method | Monomer: N-carboxy anhydride of L-leucine |
| | Synthesize an anhydride compound using L-leucine available on the market. |

Polymer: poly(L-leucine)
Obtained by polymerizing the monomer mentioned here.

| | |
|---|---|
| Solvent | Benzene |
| Characteristics | Molecular weight of poly(L-leucine) is 300,000–400,000 Da |
| | The poly(L-leucine) sponge has exceptional biocompatibility and does not biodegrade easily. |
| | Because it is a hydrophobic polymer, it is easy to release drugs in a controlled manner. |
| Gel preparation method | Gelation occurs by dissolving poly(L-leucine) in benzene solution of 70°C and returning it to room temperature. In addition, by freeze-drying it, two layers of poly(L-leucine) sponges made by a thin-film structure (outside) and a sponge structure (inside) can be obtained. |
| Uses | Used for skin treatments and open wound bandages. |
| Raw material manufacturer | Kanto Chemical, Wako Pure Chemical Industries |
| Product manufacturer | |
| Related literature | 1) *Journal of Burn Care & Rehabilitation* **12**(2): March/April (1991). |
| Chemical formula | |

$$\left[\text{NH}-\text{CH}-\text{CO}\right]_n$$
$$|$$
$$\text{CH}_2$$
$$|$$
$$\text{CH}$$
$$\text{CH}_3 \diagup \diagdown \text{CH}_3$$

polymer

$$\text{NH}_2 - \overset{\overset{\displaystyle CO_2H}{|}}{\underset{\underset{\displaystyle CH_3 \diagup \diagdown CH_3}{CH}}{C}} - H$$

monomer

---

| | |
|---|---|
| Name | Porphyran |
| Classification• network | Natural, organic, polysaccharides |
| Classification• crosslink point | Physical gel |
| Classification• fluid | Hydrogel |
| Manufacturing method | After placing *Porphyra umbilicalis* in water adjusted to pH 2 using acetic acid, it is extracted using hot water at pH 6–7. Purification is done by precipitating using ethanol. |
| Solvent | Water |
| Characteristics | Gels when cooled, but the gel strength is weaker than agarose, although it has a similar structure. |
| Gel preparation method | Cooling solution |
| Uses | |
| Raw material manufacturer | |
| Product manufacturer | |
| Related literature | 1) J. R. Nunn and M. M. von Holdt: *J. Chem. Soc.* **1957**: 1094 (1957). |
| | 2) N. S. Anderson and D. A. Rees: *J. Chem. Soc.* **1965**: 5880 (1965). |

Chemical formula

( R = H or CH₃ )

---

| | |
|---|---|
| Name | Malononitrile |
| Classification• network | Synthetic, organic |
| Classification• crosslink point | |
| Classification• fluid | Hydrogel, organogel |
| Manufacturing method | Synthesized by heating cyanoacetoamide in 1,2-dichloroethane with oxy phosphorus chloride or phosphorus pentachloride. |
| Solvent | Water, ethanol, ether, acetone, benzene |
| Characteristics | Crystal of no color, mp 31.6°C, bp 218–219°C, flashpoint 112°C, specific gravity $d_4^{20}$ (s) $= 1.1910$, $d^{35}(1) = 1.0494$, refractive index $n_D^{34.2} = 1.4139$ |
| | The interaction with concentrated hydrochloric acid makes it malonic acid. The methylene group is active and is replaced easily with sodium and bromine. |
| | Synthetic raw material for vinylidene cyanide. |
| Gel preparation method | Polymerizes by glow discharge. |
| Uses | Humidity sensor |
| Raw material manufacturer | Aldrich, Wako Pure Chemical Industries, Kanto Chemical, Nakarai Tesque, Tokyo Chemical |
| Product manufacturer | |
| Specific Notes | Toxic materials as identified by toxic material specification law in Japan. It is also called malonicnitrile or methylene cyanide. |
| Related literature | 1) B. B. Corson *et al.*: *Org. Synth., Coll.* **2**: 379 (1943). |
| | 2) B. B. Corson *et al.*: *Org. Synth.* **25**: 63 (1945). |
| | 3) E. Radeva, Kobev, and L. Spassov; Study and application of glow discharge polymer layers as humidity sensors. *Sensors and Actuators B* **8**: 21–25 (1992). |
| Chemical formula | |

$$CH_2 \overset{\diagup CN}{\underset{\diagdown CN}{}}$$

---

| | |
|---|---|
| Name | Methyl cellulose [MC] |
| Classification• network | Natural, organic, polysaccharides |
| Classification• crosslink point | Physical gel |
| Classification• fluid | Hydrogel |

| | |
|---|---|
| Manufacturing method | Obtained by reacting methylchloride at 95–100°C with alkali cellulose (pulp, linter pulp as raw material) under pressure. If the alkali concentration is high the reaction time is long. |
| Solvent | Water (a polymer with methoxy group content 25–33% dissolves in cold water but does not dissolve in hot water) |
| Characteristics | It is white or yellowish white powder and has no odor or smell, and is a noncrystalline fibrous powder. It dissolves into organic solvent as the degree of ether formation increases. It does not dissolve (mostly) in alcohol and chloroform. The solution is neutral and is stable at pH 2–12. It precipitates as a gel at high temperature. It is nonionic, so when a multivalent metallic ion is added to the solution, it does not precipitate. |
| Gel preparation method | After the solution is dispersed and swollen in warm water, when cooled to 5–10°C then a stable, clear solution is obtained. When the solution is heated to 50–70°C, the viscosity in the solution increases and gels, resulting in phase separation. This phenomenon is thermoreversible. |
| Uses | When water is added, it becomes a viscous solution. It is used to adjust the fluidity and water maintenance and plasticity of cement mortar. It is also used as a suspended polymer dispersion stabilizer, paste, cosmetics, food and medicinal compounds. |
| Raw material manufacturer | Shin-Etsu Chemical Co., Matsumoto Yushi-Seiyaku Co. |
| Product manufacturer | |
| Related literature | |
| Chemical formula | |

$R : -CH_3$ or H

| | |
|---|---|
| Name | Methyl starch |
| Classification• network | Natural, organic, polysaccharides |
| Classification• crosslink point | Physical gel |
| Classification• fluid | Hydrogel |
| Manufacturing method | Used by interacting sodium hydroxide [1] or metallic sodium [2] (see Related Literature here) with iodomethyl in liquid ammonium. Unless these reactions are repeated, complete methylation does not occur. |
| Solvent | Water, acetone, chloroform, ether |
| Characteristics | The monomer molecular weight is 190 when $-OCH_3$ content is 32.8%, and is 204 Daltons when it is 45.6%. |
| Gel preparation method | It becomes insoluble in water by adding heat or crosslinking the cellulose crosslinking agent. Crosslink or polymerize using radiation or radical initiator along with crosslinking agent or vinyl monomer. |

| | |
|---|---|
| Uses | Stabilizer of emulsion in cosmetics and medicine. |
| | Thickening agent in paper manufacturing. |
| Raw material manufacturer | Not available on the market. |
| Product manufacturer | |
| Related literature | 1) J. E. Hodge, S. A. Karjala, and G. E. Hilbert: *J. Am. Chem. Soc.* **73**: 3312 (1951). |
| | 2) K. H. Mayer, M. Wertheim, and P. Bernfeld: *Helv. Chim. Acta* **23** (1940). |
| Chemical formula | Starch–OCH$_3$ |

---

| | |
|---|---|
| Name | Laminaran |
| Classification• network | Natural, organic, polysaccharides |
| Classification• crosslink point | |
| Classification• fluid | |
| Manufacturing method | Nonwater soluble laminaran: mix *Laminaria hyperbore* in a dilute hydrochloric acid solution at 70°C, then extract and recrystallize. Soluble laminaran: after mixing *L. digitata* at low temperature with dilute hydrochloric acid solution, extract and recrystallize. |
| Solvent | Water |
| Characteristics | Insoluble laminaran $[\alpha]_D = -14.4°$ (does not dissolve in cold water, dissolves in hot water). |
| | Water-soluble laminaran $[\alpha] = -12.0°$ (dissolves in water). |
| | Sulfuric laminaran works as a blood clotting preventative. Shows an antitumor activity. |
| Gel preparation method | It gels by cooling after dissolving in hot water. |
| Uses | Blood clotting prevention. |
| Raw material manufacturer | Sigma |
| Product manufacturer | |
| Related literature | 1) V. C. Barry: *J. Chem. Soc.* **1042**: 578. |
| | 2) W. W. Hawkins and V. G. Leonard: *Can. J. Biochem. Physiol.* **36**: 161 (1958). |
| | 3) H. Saito, R. Tabeta, M. Yokoi, and T. Erata: *Bull. Chem. Soc. Jpn.* **60**: 4259 (1987). |
| Chemical formula | |

CH$_2$OH ⟨ CH$_2$OH CH$_2$OH ⟩

O O O O—CH$_2$

OH HO HO HOCH

HO HOCH

OH OH OH $_n$ HCOH

HCOH

CH$_2$OH

← Laminary biose →

Mannitol

---

| | |
|---|---|
| Name | Lichenan |
| Classification• network | Natural, organic, polysaccharides |
| Classification• crosslink point | |
| Classification• fluid | |

| | |
|---|---|
| Manufacturing method | Extracts the vegetable matter and water (except for tannin), and this precipitates when the extracted liquid is cooled. Dissolve this again in hot water, let it precipitate again, wash it with ethanol, and dry it. Use electrodialysis for manufacturing powder. |
| Solvent | Water (hot water), alkali, formamide |
| Characteristics | Dissolves in hot water, does not dissolve in cold water |
| | $[\alpha]_D = +8\text{--}10°$ (in NaOH). |
| | DP (average degree of polymerization) $= 100\text{--}200$. Has no reduction and does not color from iodine. |
| | Has white, amorphous powders that are slightly absorbent. |
| Gel preparation method | Unknown |
| Uses | Lichenes that include lichenan are used for food. |
| Raw material manufacturer | Sigma |
| Product manufacturer | Megazyme |
| Related literature | 1) N. B. Chanda, E. L. Hirst, and D. J. Manners: *J. Chem. Soc.* **1957**: 1951 (1957). |
| | 2) S. Peat, W. I. Whelan, and J. G. Roberts: *J. Chem. Soc.* **1957**: 3916 (1957). |

Chemical formula

Laminary biose    — cero biose    — cero triose

$\rightarrow \overbrace{4G_B1 \rightarrow 3Gl} \rightarrow \overbrace{4Gl \rightarrow 4Gl} \rightarrow \overbrace{3Gl \rightarrow 4Gl \rightarrow 4Gl} \rightarrow 3Gl \rightarrow 4Gl \rightarrow$

4-O-B laminary biosil glucose     3-O-B cero biosil glucose

---

| | |
|---|---|
| Name | Lentinan |
| Classification• network | Natural, organic |
| Classification• crosslink point | Physical gel |
| Classification• fluid | Hydrogel |
| Manufacturing method | Extracted from Shiitake mushrooms, separated and refined. |
| Solvent | Alkali solutions |
| Characteristics | White. No taste. No odor. A powder with no crystallization. Does not dissolve well in water. Dissolves in alkali water-soluble solutions, and dissolves relatively well in formic acid or dimethyl sulfoxide. Does not dissolve in other organic solvents. |
| | Does not show a clear boiling point. Starts carbide decomposition at $\approx150°$C. |
| Gel preparation method | Cool after heating, colloid decomposed in water. |
| Uses | Antitumor agent |
| Raw material manufacturer | |
| Product manufacturer | Ajinomoto, Morishita, Yamanouchi |
| Special note | Store in dark area. Dissolving adjuvant: Dextran 40, Mannitol-related. |
| Related literature | 1) Osaka Hospital Pharmacists Association (ed.): Survey of all medicines. |

Chemical formula

| Name | Locust bean gum |
|---|---|
| Classification• network | Natural, organic, polysaccharides |
| Classification• crosslink point | Physical gel |
| Classification• fluid | Hydrogel |
| Manufacturing method | The internal lactic part is removed from the outer seed shell and the hull of the locust seed. The removed internal lactic part is made into a fine powder, and this is dissolved in hot water, filtered, and precipitated in alcohol. |
| Solvent | Hot water |
| Characteristics | Linear chain-like polymer units have a molecular weight of 310,000. Stable in the range of pH 3–11. Viscosity 3500 cP in 1%-water-soluble solution. |
| Gel preparation method | Locust bean gum does not form gels, but when mixed with agar and carrageenan, it has the effect of improving jelly composition. In the same way as for guar gum, it forms a gel with boric acid ions under low alkaline conditions. |
| Uses | Ice cream stabilizer, cheese coagulator, bonding antioxidant for meat and sausages, frozen candy, jellies |
| Raw material manufacturer | Mayhall, FMC, Copenhagen Pectin, SBI |
| Product manufacturer | Sigma, Lone Puran, Sansho, Mitsui & Co., Yukijirushi Foods |
| Special note | Store at room temperature |
| Related literature | 1) F. Rol: *Industrial Gums*, 2nd ed., R. L. Whistler (ed), New York: Academic Press (1973), pp. 323–337. |
| | 2) I. C. M. Dea *et al.*: *J. Mol. Biol.* **68**: 153 (1972). |

Chemical formula

# INDEX

ISBN 0-12-394964-5